CROWOOD METALWORKING GUIDES

ELECTROPLATING

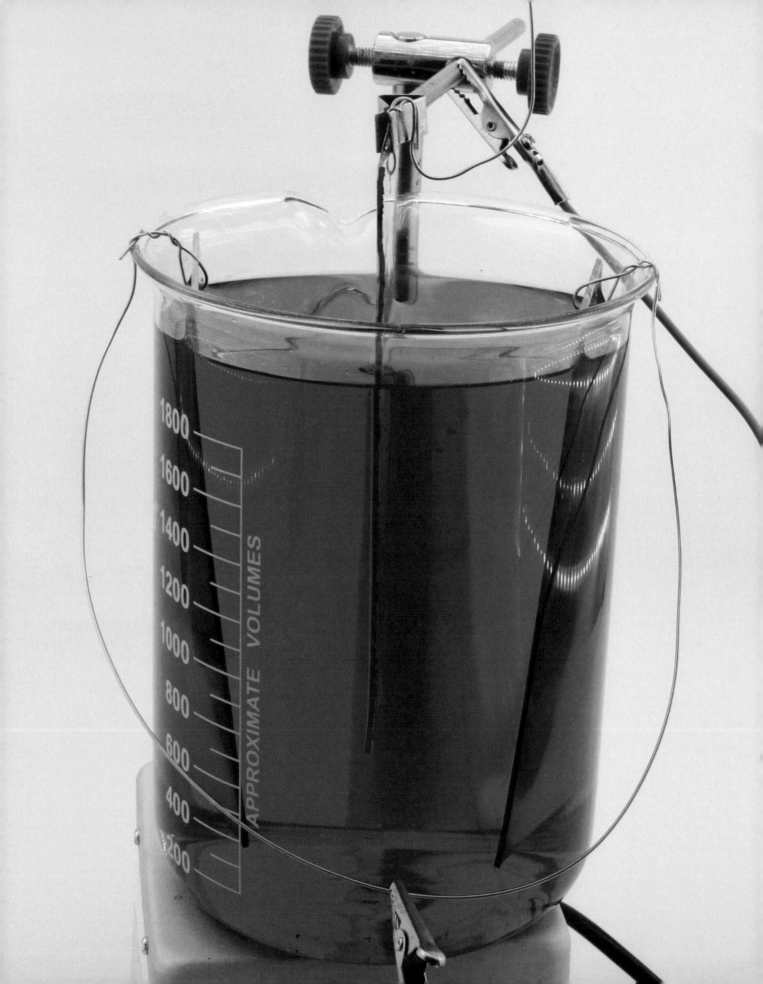

CROWOOD METALWORKING GUIDES

ELECTROPLATING

DAN HANSON AND DAVID HANSON

THE CROWOOD PRESS

First published in 2019 by
The Crowood Press Ltd
Ramsbury, Marlborough
Wiltshire SN8 2HR

enquiries@crowood.com

www.crowood.com

Paperback edition 2023

© The Crowood Press 2019

All rights reserved. No part of this publication may be reproduced or transmitted in any form or by any means, electronic or mechanical, including photocopy, recording, or any information storage and retrieval system, without permission in writing from the publishers.

British Library Cataloguing-in-Publication Data
A catalogue record for this book is available from the British Library.

ISBN 978 0 7198 4388 4

Disclaimer
Safety is of the utmost importance in every aspect of metalworking. The practical workshop procedures and the tools and equipment used in metalworking are potentially dangerous. Tools should be used in strict accordance with the manufacturer's recommended procedures and current health and safety regulations. The author and publisher cannot accept responsibility for any accident or injury caused by following the advice given in this book.

Cover design by Maggie Mellett
Typeset by Derek Doyle & Associates, Shaw Heath
Printed and bound in India by Parksons Graphics Pvt. Ltd.

Contents

Introduction

'Electroplating is the deposition of metal, from an electrolyte onto an electrode. In general, all electroplating systems are made from three elements, the anode and cathode, collectively called electrodes; the electrolyte, a water-based solution containing metal ions; and an electric circuit with a power supply to create and allow a flow of electrons. These simple elements can be brought together to create a system capable of not only restoring and decorating objects, but also make them functional and able to withstand the pressures and stresses of many engineering applications.

While many forms of electroplating exist, this book will focus on tank plating. This process can be scaled to meet the needs of large industries or to be performed on a bench in a workshop, making it ideal for the DIY enthusiast looking to restore worn and rusted parts of an old motorcycle, or to small batch jewellery manufacturers. In fact, there are countless applications in which electroplating can be employed. The earliest experiment into electroplating was performed as early as the late 1700s, where an Italian scientist plated gold onto two silver disks. Chapter 1 will cover the development of the history of electroplating as well as covering the scientific principles that underpin the process; the movement of charged particles through the system.

There are numerous formulations of electroplating electrolyte and the most common of these will be covered in Chapter 3. The most common and the most functional electrolyte is that of Watts nickel. It is an acidic solution containing nickel salts in the form of nickel chloride and nickel sulphate, with the addition of boric acid and a few other additives to aid with the brightness and uniformity of the plated nickel deposits. This nickel solution has many different variations for many specific applications. In many circumstances, more than one metal layer will be deposited onto an object. Multilayer plating is common in engineering applications or in the conservation of precious metals for the creation of jewellery. An ideal metal for an underplate is copper. Not only is an acid copper electrolyte cost-effective, it has great throwing power and conductivity, meaning that it can fill in imperfections and create an appealing finish. Copper also has great mechanical properties, ductility and thermal and electrical conductivity, making it more appealing. An electrolyte that is used a lot in the restoration of fastenings and other mechanical components is zinc. This is a favourite among home electroplaters. Its cost-effectiveness combined with its ability to give a range of finishes and the ease of use make it an ideal starting point for any electroplating enthusiast.

After safety, which will be covered in the final chapter, cleaning and preparation are the most critical steps in electroplating. Soils, the collective term for dirt, rust, oxides, oil and any other material that is not part of the base metal, is the enemy. Any soils that remain on the surface of an item when it is being electroplated will cause problems. These are most commonly seen as blistering, peeling and burning. Problems such as these as well as electrolyte and operating problems will be covered in Chapter 7. Chapter 6 will focus on cleaning itself and the various methods that can be used. The most effective cleaning method is that of electrocleaning; the use of electricity and electrolysis make the removal of unwanted oxides and paint easy. In this process, the work (the part to be cleaned/plated) is immersed in a cleaning solution and then made either the anode or cathode, positive or negative, in the electric circuit. Once current is applied, bubbles form on the surface of the metal and remove soils. For many items, multiple cleaning methods will be needed to prepare the surface ready for electroplating. It is always worth spending extra time and effort on cleaning and preparation to avoid problems and for the best finish.

Be safe. Some chemicals used can be extremely hazardous, so take extra care and precautions and always read the material safety data sheets. Also, we strongly advise that you follow the safety information provided in Chapter 9 and find additional material from the further reading listed in this book if you are still not confident. The information contained within this book is intended to give you the confidence to perform your own experiments with electroplating, from creating electrolytes safely to achieving pristine, clean surfaces and quality final pieces. Throughout the book there are text boxes with hints and tips to help with your electroplating experience; these are little bits of information we have gathered and learnt from trial and error and from the experiences of numerous home platers.

1   History and Basics

Electroplating has been developed from an art form to an almost exact science, all driven by the developing needs of industries. Aerospace and microelectronic applications of electroplating used frequently today are beyond the comprehension of the first experimenters of electrodeposition. The first experiment involving the electrodeposition of metal is a little uncertain. It is widely accepted that the first experiment was performed in 1772 by an Italian professor of experimental physics, Giovanni Battista Beccaria [1]. His experiment used the charge from a Leyden jar, an early capacitor consisting of a glass jar and layers of metal foil, to decompose metal salts and deposit metal.

The development of electroplating has closely shadowed the development of electricity production with the first published experiment in gold plating occurring five years after the creation of Alessandro Volta's, Voltaic pile. This early battery was used in the gold plating experiment by a colleague of Volta, Luigi Valentino Brugnatelli, in which the charge from the pile was used to facilitate the electrodeposition of a thin layer of gold from a solution onto two silver medals. Brugnatelli described how he had

'recently gilt in a perfect manner two large silver medals, by bringing them into communication, by means of a steel wire, with the negative pole of a Voltaic pile, and keeping them, one after the other, immersed in ammoniuret of gold newly made and well saturated.' [2]

No substantial progress was made to the development of electroplating due to the inadequacy of the Voltaic pile and other battery systems. There were, however, numerous scientists that were intrigued by electrochemical reactions (including electroplating), one of them being Michael Faraday. Among his other important discoveries and theories were his laws of electrolysis. After a few major experiments and the invention of the volta-electrometer, Faraday's results had implied two laws:

1 The chemical deposition due to flow of current through an electrolyte is directly proportional to the quantity of electricity passed through it.

2 When the same quantity of electricity is passed through several electrolytes, the mass of the substances deposited are proportional to their respective chemical equivalent or equivalent weight.

The first law can be represented by the equation $m \propto Q$, where m and Q are the chemical deposition and electric charge. It can also be written as $m = ZQ$, where Z is a constant of proportionality known as the electrochemical equivalent of a substance. The electrochemical equivalent can be expressed as

$$Z = \frac{M}{Fz}$$

where M is the molar mass of the substance, F is the Faraday constant and z is the valency number of the substance. Faraday's first law can therefore be expressed as:

$$m = \left(\frac{Q}{F}\right)\left(\frac{M}{z}\right)$$

This equation will be important later where it is used to work out the thickness or time of deposited material. These laws, along with a series of experiments, had given a sound qualitative and quantitative foundation for electroplating.

Faraday's Nomenclature

It is interesting to know that Faraday, in correspondence with William Whewell [2], established the nomenclature associated with electroplating including the terms anodes, cathode, anion, cation, electrolysis and electrolyte.

Invented in 1836 by British chemist, John Frederic Daniell, the Daniell cell [3] (a development of the Voltaic pile that allowed for a sustained, constant electrical output) provided electroplating with a suitable source of current for further experimentation, allowing for a more uniform coating of metal. Many scientists quickly took to experimenting with the new cell, one of whom was George Richards Elkington, who filed one of many Elkington patents on 'An Improved Method of Gilding Copper, Brass and Other Metals or Alloys of Metals' [4][5]. George, along with his cousin Henry, later filed many patents on electroplating. They were first interested in it as a technique to replace the hazardous gilding process, substituting these chemicals with materials that were less poisonous and easier to handle.

Later in their career, the Elkington cousins formed a partnership with John Wright, who discovered that gold and silver could be dissolved in potassium cyanide for use as an electrolyte in electroplating. A patent was hastily written and granted on the use of a cyanide electrolyte for gold and silver. The Elkingtons were granted further patents and eventually had great commercial success electroplating expensive-looking silverware and precious metals for very low cost.

Scientists in Russia, namely Moritz Hermann von Jacobi and Maximilian Herzog von Leuchtenberg, also took great interest in the Daniell cell. Initially, Jacobi repeated Daniell's experiments, finding similar copper deposits mentioned in the first experiment. He then replaced the cathode with an engraved printing plate. The engraving was removed and Jacobi was left with a metal object that had a clear impression of the plate. This is now known as an electroform. Jacobi continued his work and reported his findings to the Academy of Sciences in St Petersburg, describing it as 'galvanoplastik', translated as electroforming. Jacobi was an advisor to Leuchtenberg who went on to create the St Petersburg Electroforming Company, one of the largest at the time with more than 800 employees. The company started off by producing electroforms for printing papers, but later went on to reproduce artworks by taking electroforms of statues and sculptures.

As the Industrial Revolution expanded from Great Britain to the rest of the world, the knowledge of electroplating began to broaden. Other types of plating processes, driven by specific manufacturing and engineering applications, were adapted for commercial usage. These included bright nickel plating, brass plating, tin plating and zinc plating.

After growth of the electroplating process in the Industrial Revolution, many new material advances were made. The scientific and mathematical models were developed to help to explain how the process worked and a few improvements were made in direct current power supplies, manufacturing methods and electrolyte solutions. The cyanide gold and silver solutions changed very little until, due to a surge in the electronics industry, the hazardous cyanide baths were mostly replaced with safer acid baths[6].

Today, the growing applications for the electronics, car and aerospace industry are beginning to push the boundaries of the structure of electrodeposited metal. There have also been, justifiably, further safety regulations, laws on emissions and disposal, chemical advancements and requirements for the entire electroplating industry. Many of the chemicals that have been used for generations are now being phased out in preference for safer and less detrimental alternatives to protect workers and the environment.

WHAT MAKES A PLATING TANK?

The components of a plating tank, or in more general terms an electrolysis cell, consist of three main parts: the electrolyte, electrodes and electric circuit.

Electrolyte

The electrolyte is almost always aqueous where water is the solvent. Creation of an electrolyte solution would start with distilled, deionized (DI) or reverse osmosis (RO) water.

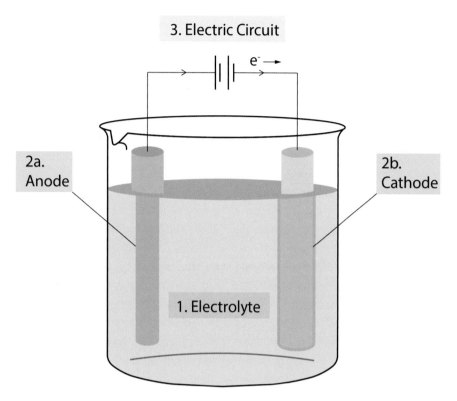

3. Electric Circuit

e⁻ →

2a. Anode

2b. Cathode

1. Electrolyte

1.1: 1. Electrolyte – the plating solution containing the dissolved metal ions is the material that connects the two electrodes. It is almost always aqueous (water-based) and has many other additives to help with the plating process. 2. Electrodes a. Anode – the positive electrode, usually made from the same metal, which is dissolved in the electrolyte and is used sacrificially. It can sometimes be an inert or insoluble material to complete the circuit. b. Cathode – this is the negative electrode and the part to be plated. It can either be a metal or another material coated with a conductive layer. 3. Electric circuit – the electric circuit consists of a source of direct current (a power supply), to deliver current to the plating tank, an instrument for the regulation of voltage and current and possibly ammeters and voltmeters to measure them.

These types of water are similar in that the processes remove ionic impurities, but they are not interchangeable. Distilled water is demineralized water that has been purified through distillation. DI water is created by running water through an electrically charged resin. The ions in the resin react with ions in the water to produce more water. DI water is quite sensitive and will start to react with air as soon as it comes into contact. Carbon dioxide from the air will dissolve and react with molecules into the water, forming hydrogen ions and bicarbonate. It also does not remove any impurities in the water, only charged particles. Diffusion is the movement of molecules from an area of high concentration to an area of low concentration until a thermodynamically favourable equilibrium is reached. Osmosis is a special case of diffusion where the molecules are water and the concentration gradient is a semipermeable membrane. The best type of water for an electrolyte would be RO water.

Ionic crystals are the next component in the creation of an electrolyte solution. These usually take the form of a metal salt lattice; *MA* is an example we will use here, where *M* is the metal and *A* is the salt. The salt can also be written as a product of two ions $M^{z+}A^{z-}$, where *z* is the valency (number of extra/missing electrons) of the ion. When mixed with water, the ion-ion bonds in the lattice of the crystal will break and the ions will interact with water to become hydrated. This can be represented by the equation:

$$M^{z+}A^{z-} + 2H_2O \rightarrow M^{z+}H_2O + A^{z-}H_2O$$

The dissolved ions become the carriers of the current in the electrolyte and are attracted toward the electrode of opposite charge to the ion itself.

Additives are another main constituent of an electrolyte. They aid with several different variables; stabilization, pH, crystal growth, metal deposition and removal and ion transportation. The main types of additives are brighteners, levellers and carriers.

Brighteners are molecules that are plating accelerators, grain refiners and/or micro-levellers. They are used to make the new metal surface look brighter; this occurs through grain refinement (the reduction of metal crystal size) and micro-levelling. Carriers work on the anode instead of the cathode. They help create and maintain a diffusion layer that regulates the flow of metal ions from the anode and, ultimately, to the cathode. Levellers are strong plating inhibitors, which co-function with brighteners to reduce metal growth at corners, peaks and edges by increasing the electrical potential (amount of work needed to move a unit of charge from one point to another) in those regions. If a metal layer is not forming uniformly on the cathode, the atoms can start to build irregular shapes (*see* Images 1.2a to d). The levellers work to even out, to level, this irregular growth and create a more uniform layer.

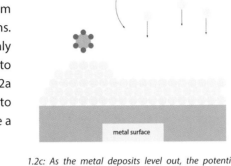

1.2c: As the metal deposits level out, the potential decreases and the brighteners are no longer attracted.

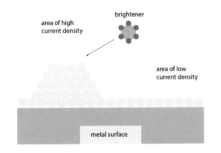

1.2a: The change in shape of the metal surface is coupled with a change in the current density (the amount of current flowing per unit area) and therefore electric potential will change in this area.

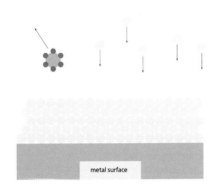

1.2d: The brighteners are free to drift away to another area of high potential.

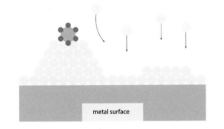

1.2b: The levellers are attracted to this change in potential and inhibit the attachment of more metal ions in this region, forcing them to be deposited elsewhere.

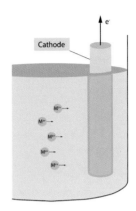

1.3: The cathode is connected to the negative of the power supply and attracts positively charged particles called cations. These particles are the metal ions, M^{z+}, that react at the cathode to form metal atoms and other products to create a new surface layer in a reaction called reduction.

replenishment these ions when they are deposited from the solution onto the cathode. In some systems, inert anodes, such as carbon and platinized titanium (titanium anodes with a thin layer of platinum on the surface), are used.

Cathode

The cathode, as mentioned above, is the part to be plated, and is often called the 'work'. Prior to electroplating, the cathode will have gone through a rigorous cleaning process to prepare the surface, removing soils and impurities and activating the base metal.

Anode

The anode is the electrode that is connected to the positive end of the circuit. It is made from the same metal as the ions in the solution and is used to

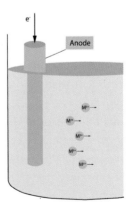

1.4: As the metal ions in the electrolyte are used up, more ions are stripped from the anode in a process called oxidation. In this process electrons are lost, and metal ions are created. Occasionally, inert anodes may be used.

The External Circuit

The most important part of the external circuit is the power supply. In all applications of electroplating, this must be able to deliver direct current, DC. The ions in the electrolyte move due to the electric current (and therefore electric field) supplied by the power supply. The metal ions (cations) flow with the electric field, toward the cathode, and the salt ions (anions) flow in the direction of the electrons, against the electric field, toward the anode. For a metal layer to be formed there must be a net migration of metal ions in the solution toward the cathode. With an alternating current, there will be no net ion flow; the same number of ions will join the cathode as the amount that leaves the cathode. This is because the electric field direction is oscillating over time. There is a plating technique called periodic current reversal that may seem comparable to using an alternating current, however, the net migration is still high as the amount of time the current is reversed is proportionally much smaller than when it is in the correct direction.

As part of the circuit, there must be an instrument that will be able to control the current. This is called a resistor, variable resistor or rheostat. A variable resistor is usually a coil of wire wrapped around an insulator with a wiper that can move down the length of the coil to change the resistance and therefore current supplied to the anodes. It is useful to also have ammeters and voltmeters to measure the current and voltage respectively.

HOW IT WORKS

Electrodeposition can be summarized in three main steps:

1 Ionic migration – which happens in the electrolyte solution
2. Electron transfer – which occurs at the metal-solution interphase
3. Incorporation – which ends on the surface of the metal

When a metal ion is in an electrolyte with an applied current, it will migrate toward the cathode due to its positive charge. The ion will then attach to the surface of the cathode via the transfer of electrons and become incorporated into the metal to become part of an electroplated layer. These steps of electrodeposition can be summarized as the reduction (the gaining of electrons) of metal ions from an electrolyte to form metal atoms. The reduction reaction takes the form of this equation:

$$M^{z+} + ze^- \rightarrow M$$

where M^{z+} is the metal ion with number of charges, z, which is equal to the number of electrons e^-, and M is the metal atom. This reaction can happen in two different processes, electro deposition powered by an external power source or electroless deposition. Electroless deposition, or autocatalytic deposition, involves no power supply and will be explained later. For now, we will focus on the three steps of reduction.

Ionic Migration and Electrolyte Solution

The basis of most electrolytes is water. Image 1.5 shows the constituents of the molecule as well as the locations of the atoms. The dipole moment of water molecules allows them to bond weakly with surrounding atoms, ions and molecules. When water bonds with more water molecules, water clusters are formed. These can be seen in Image 1.6. The larger and higher number of water clusters contained within a solution decreases the ability of water to conduct electricity. Due to the ionic dissociation of water (the breaking of water molecules to form hydrogen and hydroxide ions), there will always be ions to reduce the number of water clusters. The more ions in a solution, the lower the dielectric constant of water will be and the better it will be to conduct electricity. Water clusters can be broken up further with the addition of metal ions. When metal ions are dissolved into water, water molecules bond to the ions and form a hydration shell. The ion and shell then behave as one entity. When current is applied, these hydrated ions move in the direction of the current due to the electric field created.

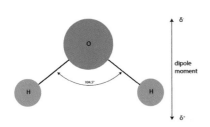

1.5: *Water is made from two hydrogen atoms and one oxygen atom, hence H$_2$O. The two hydrogen atoms form an isosceles triangle of covalent bonds with the oxygen atom. The arrangement of the atoms in the molecule gives rise to a permanent dipole moment. A dipole is the separation of two charges of equal magnitude. The side with the hydrogen atoms is positive and the side with the oxygen atom is negative. This arises due to the way in which the electrons are shared between the atoms. Water can therefore bond weakly to other charged particles, ions and other water molecules, for example.*

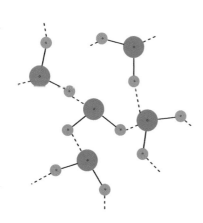

1.6: *Water clusters form due to the dipole moment of the water molecule. The attraction force is shown by the dashed line between the positive hydrogen atoms and the negative oxygen.*

Metal-Solution Interphase

An interphase is a region between two phases (types of matter) that have a different composition. In an electroplat-ing cell, the interphase is the region where the anode/cathode and electro-lyte meet. We have already talked about how, in the electrolyte, the water atoms can be attracted due to their dipole, so now we will talk about the structure of the metal surface.

When it comes to electroplating, the metal surface of the cathode is the most important feature. The surface usually refers to the first layer of atoms, or the first few layers. In regular circum-stances, the metal surface is relaxed. This term is used to express the way in which the first layer of atoms has a smaller bond strength than those in the rest of the material. This is because they are not surrounded by other metal atoms. When a metal is placed into a solution containing ions of the same metal, there will be a natural exchange of ions between the two phases, even with no current involved. This can be seen in Image 1.7. Some ions enter the solution from the relaxed metal surface and some enter the metal surface from the solution until a natural equilibrium is reached. This equilibrium can be rep-resented by the equation:

$$M^{z+} + ze^- \leftrightarrow M$$

The presence of extra electrons slightly charges the metal surface and interphase, which attracts water mol-ecules. The water has an increased con-centration in the interphase and some bond to the surface of the metal.

Incorporation and Electrodeposition

Bringing together the previous sec-tions, we will describe how the metal ions are incorporated into the metal surface and how a new layer is formed. When an external power source is applied to the circuit, the hydrated metal ions begin to move in the solu-tion. Their charge carries them toward the cathode, through the interphase and onto the surface of the metal. As the ion moves toward to the surface, it slowly begins to lose water mol-ecules, becoming less hydrated. When it reaches the surface, the hydrated ion reaches a half crystal position. This is where the ion loses half of its water molecules and becomes bonded to the lattice with half of the bonding energy available. This allows for diffusion of the ion along the surface to find a point in the lattice where it can have the lowest energy state; where it is the most stable. This point is known as a kink site. Here, the ion gains electrons and is incorpo-rated into the surface.

Lattice and Crystal Growth

As more ions follow the above process, the surface grows. It can grow in two different ways, through layer growth or crystal growth. Layer growth is the spreading of steps, or discrete layers, along the surface of the crystal. Crystal-lite growth occurs when crystallites are formed on the surface and are grown outwards. A coherent metal layer is formed as a consequence of coales-cence of these crystallites.

ELECTROPLATING OF ALLOYS

The electroplating of alloys is the codeposition of two or more metal ions onto the metal lattice. It works in

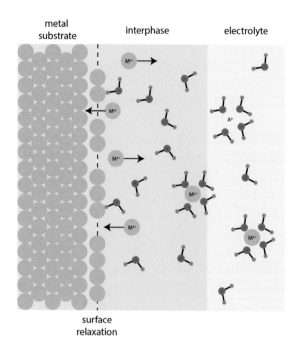

metal
substrate

interphase

electrolyte

surface
relaxation

1.7: The exchange of ions occurs due to the reduction of hydrated metal ions close to the surface of the metal and due to oxidation. Oxidation is the opposite of reduction, electrons are lost from a metal atom to form a metal ion and electron.

the same process as above for a single metal. The hydrated ions migrate toward the cathode where electrons are transferred, and the ions reach the surface and then are incorporated at growth sites. Alloys can often be more useful in terms of finished parts as their crystal deposits have properties that cannot be achieved through plating single metals. They can be denser, harder, more resistive to corrosion, tough, stronger and much more. When plating with alloys, there is one condition that must be met; the metals must behave, atomically, in similar ways.

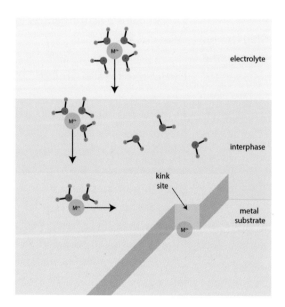

electrolyte

interphase

kink
site

metal
substrate

1.8: A kink site is a gap, break or ledge in the lattice that can give the lowest energy state. The ion diffuses across the surface until it reaches the kink site, where it will gain electrons to become a metal atom that is incorporated into the metal lattice.

2   Types of Plating

TANK PLATING

Tank plating accounts for almost all electroplating processes used in industry today. Its history, development and process have been detailed in the previous chapter. In this section, different types of tank plating will be covered; manual, barrel and rack, as well as multistage electroplating processes such as multi-layer plating, periodic reverse current electroplating, as well as why they are used and examples of their end use applications.

RACK PLATING

Rack plating, along with barrel plating, makes up the major proportion of industrial plating processes. This method of plating is both space and time efficient as many parts are placed onto one conducting rack.

The spine of the plating rack is the backbone supporting each part, it must therefore have enough strength to not only carry the weight of all the parts in the tank but also to lift and transport them. The spine must also be wear-resistant and must withstand the applied forces of prolonged usage. Usually, plating rack spines are made from copper covered in an insulating material. This material is a liquid resin, such as PVC, that is painted onto the

2.1: A simple rack used to zinc plate steel bolts. The spine is made from insulated copper with copper arms.

part and then cured. Primarily, this is to help with corrosion resistance and to make the rack non-conductive where immersed in solution – if it had no resin coating then the rack would be plated over and over. Copper is used due to its low cost relative to its high conductivity. Using a less conducting material, like steel, would lead to lower electrical efficiency due to resistance-induced heat loss. Steel can be used in situations where more support is needed.

In some circumstances, the plating rack may not have arms at all and the tips simply extrude from the spine. In other cases, where extra length and strength is needed, arms are added to the spine.

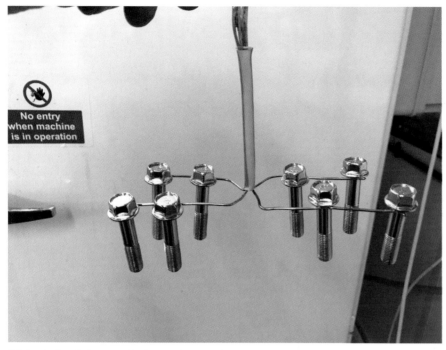

2.2: *A variation of a home-made plating rack, again, the spine is made from insulated copper and copper arms hold bolts for plating.*

2.3: *Final plating rack variation in which the spine is attached to a bus bar that is placed across the length of the plating tank.*

able; each has its advantage. Fixed tips are less expensive when first creating the rack and they can be very strong as they can be incorporated into the spine. The major problem with fixed tips is longevity. If you have a rack with ten tips and one breaks, then the amount you can plate at any time has dropped by 10%. This is where replaceable tips come in. While they may be more expensive to have, they offer complete replaceability, keeping the rack output at 100%. They also help increase the usability of the rack, allowing parts to be added and removed from the rack by removing the tip. Replaceable tips also allow for the use of different tip designs to suit the need of the parts you want to plate.

While large racks are necessary for industrial electroplating that requires a large turnover of parts, the fundamentals of rack design can be used to create small, bespoke racks for home use. As a home plater, most of the parts that you will plate will be different from each other – they will have different geometries and will therefore need to be orientated differently within the plating tank to overcome problems with low and high current density. So, each rack may have to be different. When making your own rack, there are considerations to be made:

- What kind of plating will be done?
- How many parts need plating?
- To what solutions will the rack be exposed?
- What tips do I need?
- How should the part be orientated to achieve the best finish?
- Will this design provide quick and easy racking and unracking?

Plating rack tips are the part of the rack that connect to the cathode, to the item to be plated. The two most common materials used for rack tips used in the electroplating industry are phospho- rized brass and stainless steel. Occasionally titanium will be used but this is only in special circumstances due to the larger electrical resistance of this metal. The tips can either be fixed or replace-

An ideal rack will be the results of these answered questions.

BARREL PLATING

Barrel plating, as mentioned, is one of the two main industrial plating processes that are used in electroplating. Like rack plating, barrel plating can be replicated on a smaller scale for home use. Essentially, barrel plating comprises of a rotating barrel containing all the parts to be plated. The barrel itself is placed in sequential tanks containing cleaning chemicals, rinse solutions and plating solutions. The inside of the barrel contains numerous electrical contacts that act as cathodes. While tumbling, the various parts inside the barrel contact the cathode. These parts then act as electrical connectors themselves by conducting electricity to other items within the barrel, thus increasing the efficiency of the whole process as the entire surface of all the parts become the cathode.

The advantages of barrel plating are:

♦ Speed and efficiency: a large volume can be plated at a single time.
♦ Compact design leads to space savings and smaller tanks.
♦ Labour efficient: the work remains in the barrel as you move the barrel from one solution to the next with minimal handling compared to rack plating.
♦ Versatile: if the work can fit in the barrel door then it can be plated.
♦ No special formed anodes, auxiliary anodes, shields, etc.
♦ Tumbling produces a more uniform

2.4: *An example of early barrel plating systems from the 1920s book,* The Modern Electroplater *[20].*

plated finish due to the changing contact points.
♦ No agitation in the tank is needed – leading to homogenous solutions.

The main use of barrel plating is for the final corrosion protection finishing, usually on fastenings, that do not require any handling or further finishing. Barrel plating is not usually used for decorative or engineering finishes due to the tumbling of parts, which is inherent to the process.

MULTISTAGE ELECTROPLATING

Multistage electroplating is the electrodeposition of more than one layer of metal. In most circumstances, more than one layer is needed. An example of this is when plating onto a zinc substrate. A strike followed by a copper layer followed by a bright nickel layer is common.

Each different metal has its own purpose and properties, which lead to improved strength and corrosion resistance. A layered combination of different metals can drastically increase many of the corrosion and mechanical properties, often adding up to be better than the sum of its parts. There are a vast amount of combinations possible and most with specific purposes, be it for increased strength for aerospace parts, cost reduction for precious metal jewellery plating or for increased corrosion protection for outdoor structures. The stages that can make up most multistage electroplating applications are:

♦ **Initial strike plate** – the main role of this stage is to achieve a good, adherent bond to the work. In plating aluminium, an electroless nickel strike is used before other metals can be deposited on top.

◆ **First metal layer** – this can be a wide range of metals depending on the application and is used as a thick plate to fill holes and cover imperfections. For many applications, copper is the first metal to be deposited due to its high conductivity, its micro-throwing power and other mechanical properties.

◆ **Second metal layer** – a second layer will usually be used to achieve a mirror bright finish and is often the same thickness of the previous layer. Nickel is often used in this layer due to its excellent corrosion resistance and the ease at which it will form a lustrous finish.

◆ **Final layer** – this final stage is usually coupled with the passivation stage (*see* Chapter 8). An electroless chromate conversion coating is the most common final strike plate due to the final decorative finish and corrosion protection. This final layer can also be other precious metals, such as gold or silver, which due to cost and material type do not need a thick layer.

Strike Plating

A strike plate is thin layer applied to a surface that creates a well-adhered coating on the substrate, onto which further plating can be added. It can also be used as a final decorative finish. The thickness of a strike plate is very thin, usually 0.5–5 microns.

Strike plates are used for a variety of purposes; nickel strike prior to chromium plating, nickel strike prior to gold/silver plating, and zincate strike for the electroplating of aluminium.

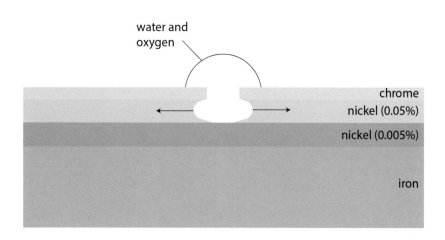

2.5: The multi-layered nickel works so well due to the two types of electroplated nickel, and more specifically, their sulphur contents. The first layer is a semi-bright nickel and would have a low sulphur content, usually around 0.005%. The second nickel layer, a bright nickel layer, would have an order of magnitude higher concentration of sulphur, 0.05%. This difference in concentration makes the semi-bright nickel layer more noble than the bright nickel above it. Moisture or atmospheric contaminants make their way into the metal, through cracks and micropores, first penetrating the chromium layer. As chromium is more noble than bright nickel, the corrosion will spread further into the bright nickel layer. On the interface between the bright and semi-bright layers, the corrosion will begin to move laterally. Again, this is due to the change in nobility due to the sulphur. Essentially, the three layers create a nobility sandwich. The result is that the bright nickel will form flat-bottomed corrosion pits and will eventually corrode fully with very minimal effect to the substrate underneath. This combination can work with various metals and is relatively easy to perform at home. Simply check the nobility of the metals/alloys you want to plate and make sure the lowest nobility is in the middle.

Most strikes are nickel, because it can be deposited onto most metals and materials with high levels of adhesion. This strike mix is most commonly an acid chloride nickel solution, often referred to as a Wood's nickel strike. Further information on this formulation can be found in Chapter 3.

Electroplating onto aluminium can only be done effectively using a strike plate. Immersion deposition of a strike layer is the widest-used technique for plating onto aluminium and its alloys. The immersion solutions are zincate- or stannate-based. They prepare the surface for electroplating by first etching the surface, stripping the oxide layer from the aluminium. Then, a thin film of either zinc or tin (depending on the solution) is deposited before the re-formation of the oxide layer. The adhesion of this film, that is the bond strength between the aluminium and zinc/tin, can be equal to or greater than the tensile strength of the aluminium, giving great final qualities of the final electroplate. After the zincate/stannate, a strike coat is often applied. This strike plate can be copper, nickel or electroless nickel.

Multilayering

As mentioned, multilayering improves many aspects of an item and gives a better overall finish for both decorative and practical applications. A very common example of a multilayer electroplating process is triple-layering involving the plating of dull nickel, bright nickel and chromium. Multi-layered nickel coatings outperform single layer coatings of the same thickness and are usually employed for objects exposed to severely corrosive environments. In combination with a chromium strike layer, these multi-layered coatings have outstanding corrosion resistance for a long period of time. Specific examples where this process is used are on exposed metalwork close to the shoreline where the wind and salt spray make for a very corrosive atmosphere.

Periodic Current Reversal

Periodic current reversal electroplating, or PCR for short, is the most common of numerous current modulation methods of electroplating. It begins in much the same way as regular electroplating, using the same set-up, chemicals and variables. Once plating begins the current is periodically reversed. When the current and time are plotted on a graph, a digital waveform is formed. The advantage of using PCR is that it increases current efficiency and improves crystallization due to interstitial hydrogen being oxidized during the reverse current cycle. This improves throwing power and can reduce the roughness of the deposit by up to 25%.

2.6: The typical waveform of periodic current reversal plating cycle. It is organized into anodic and cathodic sections. Initially the current is cathodic with a magnitude of I_c amps which continues for a time of t_c seconds. The current then switches to anodic with a magnitude of I_a amps, which is the reciprocal of I_c. The length of time that the current is anodic is much smaller and is represented as t_a. This wave repeats continually until the end of plating.

BRUSH PLATING

'Swab plating' and 'spot plating' were terms, along with a few others, that described the DILAC process of selective electroplating, now known more commonly as brush plating. Brush plating was designed by J.G. Icxi of Laboratories DALIC, Paris [7]. It is a conventional electroplating process, employing the same fundamentals as those described in Chapter 1. With brush plating, however, no large tanks are needed as the work is not immersed in the electrolyte. Rather, a brush with a porous tip is used to apply the electroplating solution to a specific area. This process is used to electroplate materials accurately, offering increased portability, flexibility and ease over conventional tank plating using simple and compact equipment.

The requirements of brush plating are a brush, small power supply and very small volume of electrolyte. It can be made portable, meaning that it is not confined to a shed, garage or workshop and can be applied 'in the field'. The areas in which tank plating are weak, brush plating fills. Brush plating can be used to cover small areas easily, by masking or taping areas that do not need plating. It can be used to plate items that are too large or have difficult geometries for tank plating. The deposit thickness can be easily controlled with

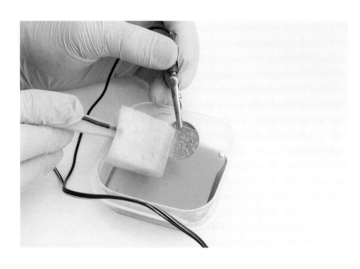

2.7: *A typical brush plating set-up used to plate bright Watts nickel. While brush plating can be used to restore small areas on large objects, it is also excellent at covering small objects very quickly.*

in the solution begin to form a new surface layer.

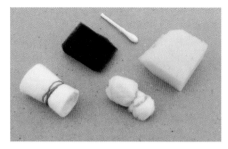

2.9: *These materials are suitable for brush plating tips. They are from left to right; 5μm PP cloth, 100μm PP foam, cotton bud, cotton wool, and 50μm PU sponge.*

brush plating and will usually not need any mechanical finishing such as buffing and polishing. Another major benefit is that it can be done at room temperature.

Brush Plating Brush

The core of the brush is an anode. Like tank plating, this anode can be made from an inert material, carbon, titanium, stainless steel, or from the same metal that is being deposited from the solution. That is, if an item was being plated with nickel, then the core of the brush can be nickel also. In this way, nickel ions will be released from the nickel core and will migrate into the solution, replenishing the metal con-

centration and allowing the solution to last longer. The core will need replacing if it is sacrificial.

The main body of the brush can be made from almost anything if it is not attacked by any solution that it contacts. Mostly, the brushes are of an ergonomic shape, like a paint brush, making application and use easier for prolonged periods.

The tip of the brush is made from a porous material, often foam, cotton wool or other similar materials, which are soft and allow the electrolyte to be absorbed and contained within. The tip also covers the anode so that a short is not created. When the saturated tip of the brush contacts the substrate, a circuit is completed and the metal ions

Method

The methodology for brush plating will be described here as the later sections in the book concern tank plating only. Before any plating – brush or tank – the part should be cleaned thoroughly. A detailed explanation is contained within Chapter 6. Once the part has been cleaned sufficiently, brush plating can begin.

The way in which an electroplating brush should be used is like that of a paint brush; continuous strokes. First, the brush needs to be connected to the positive output of a power supply and the work needs to be connected to the negative. This power supply can be very small, a 600mA variable voltage power supply will often suffice. In any case, the power supply needs to provide a direct current.

The tip of the brush is then submersed in the electrolyte, absorbing the solution, which is then applied to the surface of the part to be plated and moved at a constant velocity (speed and direction). Touching the tip to the

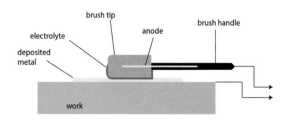

2.8: *Diagram of components of a brush plating system: tip, handle, anode core and solution.*

2.10a: Constant motion over the same section of work will deposit metal from the solution. This is the work, a piece of copper sheet, before any plating.

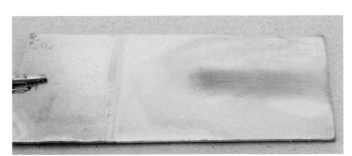

2.10b: The work after one brush stroke of black nickel solution.

2.10c: The work after three strokes.

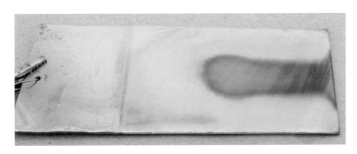

2.10d: The work after ten strokes.

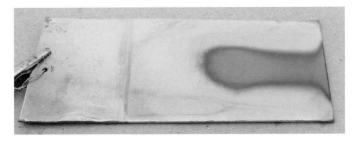

2.10e: The work after twenty strokes. The new layer looks visible from the first application, however, increasing the number of stokes increases the thickness of the deposit.

object completes the circuit, allowing the free flow of electrons and thus electroplating. The deposition process now follows the same format as per Chapter 1. If the brush moves too fast, then minimal plating will occur. If the brush is moved too slow or does not move at all while in contact with the substrate, then burnout can begin to occur. Brush plating should be performed over a container to collect the solution that drops from the item. This can be reused and plated out until the levels of metal ion have dropped and need replacing.

Operating Conditions

◆ Between 2-8 Volts
◆ Maximum 3 Amps
◆ Solution temperature from 18°C to 60°C
◆ Make up of solutions can be found in Chapter 3 – regular tank plating electrolytes can be used.

Applications

As mentioned earlier, brush plating can cover the deficiencies and limitations of tank plating. This can be either to plate objects with difficult geometries; that are either too large for a tank or have internals that cannot be plated with shields or auxiliary anodes. Brush plating can also be used to repair areas of a larger part. This can save time by only cleaning and plating a small area rather than the whole object. Brush plating as a process is not widespread when compared to tank plating. Most applications involve repair and restoration outside the plating workshop and the recovery of valuable/irreplaceable components.

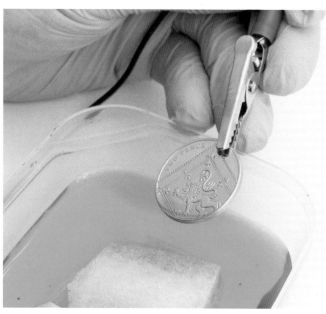

2.11: *Dark deposits will grow around the edge of the brush tip, in areas of high current density. This can also occur at boundaries or edges of the work and can be seen as the shadowing in the bottom edge of the coin. If this happens, increase the speed of movement or decrease the amperage supplied to the brush.*

available with which the ions can react. The two reaction equations are below.

Electroplating:
$$M^{z+} + ze^- \rightarrow M$$

Electroless Plating:
$$M^{z+} + Red \rightarrow M + Ox$$

The difference in the equation is *Red* and *Ox*. *Red* stands for reducing agent. These are molecules that take the place of the source of electrons in the reaction. *Ox* are the oxidation products. Overall, the metal ions react with the reducing agents on a catalytic surface. The ions take electrons from the reducing agent, and as a result metal atoms and oxidization products are formed.

There are several advantages to using electroless plating over electroplating, the main one of which can be seen in Image 2.13. Electroless plating can achieve a uniform coating on almost any surface.

Anodising

Anodising can also be achieved using a brush. While not technically plating, rather expansion of an oxide layer, the process can be critical in the repair of expensive and/or irreplaceable machinery. The process occurs in the same way as brush plating, the anodising solution is absorbed by the brush tip, which is then applied to the object to be anodized.

There is no electric circuit and there is only one electrode. The electroless reaction takes a slightly different form from the reaction described in Chapter 1, this is because no electrons are freely

2.12: *A simple diagram of an electrolytic cell for electroless. The basic components are water, metal salts, electrode with catalytic surface and a reducing agent. The reducing agent acts as the source of electrons that allows the deposition of metal on the electrode to occur.*

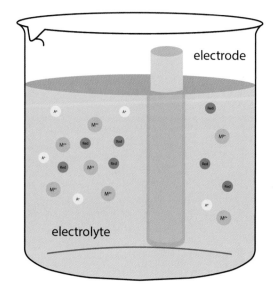

ELECTROLESS PLATING

An electroless plating tank may seem simpler than an electroplating tank.

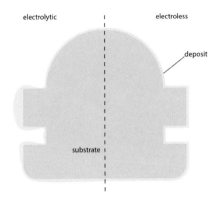

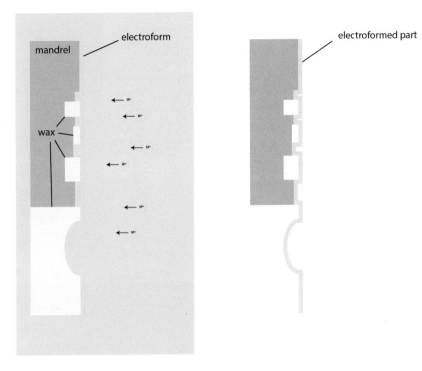

2.13: When an item is the cathode of an electroplating circuit, it will produce varying current densities throughout its complex geometries. Edges, peaks and corners will have a high current density. This is because the metal ions can reach that part more easily – it is the path of least resistance. Areas with a high current density will receive a thicker deposit as the metal will prefer to attach here as it is easier. For holes, hollows and channels, the current density can be much lower. Ultimately, these areas do not receive as much current and therefore the deposit is much thinner. This can be seen in the left-hand side of the diagram. While the current density of an object can be accounted for by using addition agents (such as levellers), shields, anode distance and auxiliary anodes, there will still be a non uniform coating. This problem is inherent to electroplated parts; for most applications, however, this variation does not cause an issue. As there is no external supply of current in electroless plating, the issue of current density does not play a factor and, therefore, a uniform layer is achieved. This is due to the localized reduction of the metal ions and oxidation of the reducing agent. Overall, this means that even the most complex of geometries may be plated if the surface is activated and catalytic.

2.14: Diagram showing the electroforming process with a mandrel containing channels filled with wax, which are plated over and then removed.

Surface Activation

For electroless deposition to work, the surface of the substrate must be catalytic; when the surface contacts the solution, a chemical reaction takes place because of the catalytic nuclei of the surface. There are several metals, known as transition metals, which lie in the groups 4–11 on the periodic table, which will have catalytic surfaces after being properly cleaned and etched. Noncatalytic surfaces, such as non-conductors, noncatalytic metals and noncatalytic semiconductors, must be activated prior to electroless plating. To produce a catalytic surface, electrochemical methods are usually employed – such as electrolysis. Photochemical methods, where light is used to reduce atoms to create catalytic nuclei on the surface, can also be employed.

Electrochemical Activation

In electrochemical activation, metal catalytic nuclei are produced on the surface through a similar reduction reaction to electroless plating. The metallic ion reacts with a reducing agent to produce metal atoms and oxidation products. The most common catalyst and reducing agents are palladium *Pd*, and tin *Sn*. The overall activation can be written as:

$$Pd^{2+} + Sn^{2+} \rightarrow Pd + Sn^{4+}$$

After this occurs, palladium atoms are adhered to the surface to form the catalytic material that is needed for the following electroless deposition.

Electroforming

As mentioned in Chapter 1, electroforming was first conceived by the Russian scientist Jacobi. In electroforming, parts are formed using electrodeposition. A layer of metal is electrodeposited on a material that is removed from the deposited metal once the required thickness has been achieved. The part to be plated in electroforming is called a *mandrel* or *mould* and the metal that is separated from this is the final part and desired product. The mandrel is often a negative of that part to be created. With some types of electroforming it is important that the deposited metal is non-adherent, i.e. does not stick, to the mandrel.

The typical electroforming process, for industrial applications such as aerospace, involves the filling of machined channels with wax or other low-melt materials, plating with copper to seal the channels and then removing the wax. The process can also involve the creation of a complete component, such as the copper cylinders in Images 2.15 and 2.16.

Electrofabrication is a similar process, however, the mandrel becomes an integral part of the final product. Often in electrofabrication, only certain areas are electroplated selectively using auxiliary anodes or shields.

For electroforming, solutions must have a high *throwing power*. This term describes the ability to deposit metal

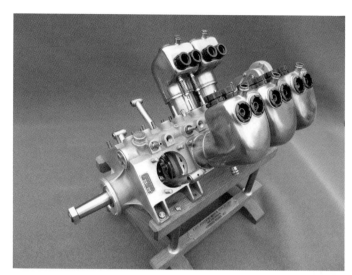

2.15: A half-scale V8 aero engine replica in which each cylinder is covered with an electroformed copper jacket that creates the space for cooling water.

2.16: The cylinders are made of a tin-bismuth low melt alloy that was subsequently melted out to leave the water space. The jackets bond with the various flanges and are permanent. In the photo you can see a bare cylinder with a covered one each side. The thickness is around .025in on the flat surfaces, and thicker around tight curves such as the inlet stub pipes.

evenly in varying current density areas. It is controlled by the relationship between the factors that influence current distribution. There are numerous factors that affect current, and therefore metal distribution: object geometry, cathode polarization, cathode efficiency-current density relationship and conductivity. An electrolyte with a high throwing power can deposit a coating that is uniform – so has almost equal thickness across the entire surface – on both recessed and prominent areas.

3 Tank Plating: Material Composition

There are a very large number of electroplating solution formulations available, and with the addition of the vast number of proprietary additives this adds up to an overwhelming quantity. In this chapter, we have outlined some commonly used, inexpensive and relatively safe formulations, listing their constituents and operating conditions so that you can create them yourself. They are listed in alphabetical order for ease of use.

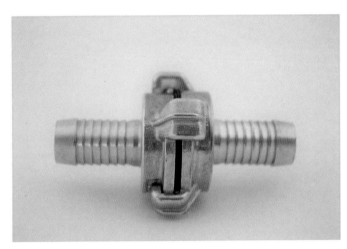

3.1: A typical electroplated brass finish.

The Right Water

In all solutions, purified water should be used. RO water is ideal but deionized or distilled are adequate.

BRASS

Brass plating is one of the oldest plating processes after the discovery of copper plating. The addition of zinc has been employed to plate the alloy. All brass alloys can be plated, ranging from pure copper to pure zinc. Brass plating is usually used for decorative finishes, with strike plates covering bright nickel layers. Thicker layers of brass can be deposited, but they tend to form nodules and larger grains with extended plating times. This is due to the lack of available additives. When followed by buffing and polishing, thicker plates can be used to fill pits and restore older brass parts. There are a range of post-plating processes available to brass to give it a unique, antique look. These include burnishing, tumbling, patinas and conversion coatings.

Most commercial brass plating is done with a cyanide solution. This is

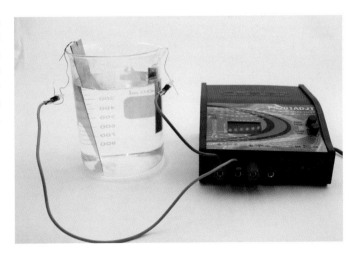

3.2a: Primed brass solution. The initial set-up of the priming solution. The sodium hydroxide solution looks clear and the anodes are clean.

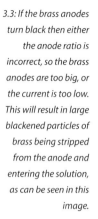

3.3: If the brass anodes turn black then either the anode ratio is incorrect, so the brass anodes are too big, or the current is too low. This will result in large blackened particles of brass being stripped from the anode and entering the solution, as can be seen in this image.

3.2b: When the priming process begins, hydrogen and oxygen are formed, creating bubbles and turning the solution cloudy.

3.2c: As the concentration of copper and zinc increases, the solution begins to turn blue.

because the cyanide complexes of both copper and zinc are very stable and their respective potentials are very close, meaning that codeposition

is very easy. As cyanide solutions can be very dangerous, we will focus on the non-cyanide solutions. These have had small commercial success, but for home plating they offer an inexpensive, safer and highly effective way to plate brass on a small scale. Some of the non-cyanide solutions that can be used are pyrophosphate solutions [8] and EDTA (ethylenediaminetetraacetic acid, disodium salt)/glycerol [8] [9] solutions.

Brass plating can also be achieved through the simple use of brass anodes and a primed sodium hydroxide solution. It is important when

3.2d: The final brass solution should be a dark but translucent colour and the steel anode should have a clean layer of brass.

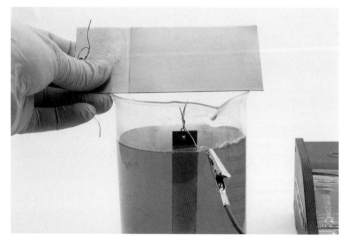

using a simple sodium hydroxide brass solution to prime it correctly. This is achieved by plating a cathode using a brass anode that is proportionally smaller. To create one litre of primed sodium hydroxide solution, mix one litre of distilled or deionized water with 110g of sodium hydroxide and stir until fully dissolved. Make sure to do this in a well-ventilated area as fumes are given off when mixing and priming a corrosive. For the cathode, use a cleaned prepared piece of mild steel or copper measuring $12.5 \times 2.5 \times 0.15$cm . The thin brass anode should measure $2 \times 2 \times 0.2$cm . The temperature of the electrolyte should be 20°C and the current should be set at one amp. Run the priming process for an hour, during which time the solution will change from clear to blue. Once the solution has been primed it is ready to use in a larger system.

When plating using a simple sodium hydroxide solution, operate at *1.5 – 2.5 A dm⁻²*. The anodes should always look clean and appear brass-coloured during plating.

This method of brass plating is for

3.4: If you plate with black-looking brass anodes the balance between copper and zinc will alter dramatically, resulting in an alloy with little copper.

from a light copper (pink bronze) through light and dark gold, to a silver colour (white bronze). Alloys with tin ratios between 10–12% have a lustrous gold colour, making it an inexpensive alternative to gold plating for costume and non-precious jewellery. As bronze is less porous than copper, it can often replace copper in circumstances that require increased corrosion resistance. Similarly, bronze with a higher percentage of tin, around 20%, can replace nickel in multi-layered decorative finishes. White bronze is now being used as a barrier plate instead of nickel on jewellery due to increasing allergic reactions to nickel. White bronze contains 55% copper, 30% tin and 15% zinc.

Again, comparable to brass, the most common bronze plating electrolyte is cyanide-based. While this produces great deposits, the environmental and health impacts owing to its high toxicity mean that we will not consider it any further in this book as other solutions are available.

The most common, and most studied, bronze solution is simply a sulphate copper solution with the addition of 5–10% tin salts to the electrolyte. These salts must be replenished regularly as the only anodes in the tank are copper. The major drawback from this type of solution is the spontaneous oxidation of the tin ions, forming particles that drop to the bottom of the tank and cause a loss in overall usable tin in the electrolyte. Pyrophosphate copper electrolytes with added tin salts have also been used. They also have a similar problem to sulphuric-based solutions and need balancing continually by adding extra salts.

	Formula	Electrolyte Composition (grams per litre - g L⁻¹)		
		Pyrophosphate	EDTA	Simple Sodium Hydroxide
Copper Sulphate	$CuSO_4$	5	30	
Zinc Sulphate	$ZnSO_4$	50	30	
Potassium Pyrophosphate	$K_4P_2O_7$	300		
Sodium Hydroxide	$NaOH$		120	110 + primed solution
EDTA/Glycerol			40	
Operating Conditions				
Temperature (°C)		30–60	20	20
Agitation, Filtration		Air or mechanical, coarse filter media	Air or mechanical, coarse filter media	Air or mechanical, coarse filter media
Cathode Current Density		50 A m⁻²	50 A m⁻²	50 A m⁻²
Anodes		Brass (same ratio as solution)	Brass (same ratio as solution)	Brass (the area of brass anode to cathode is 1:8)
pH		7.5–8.5	7.5–8.5	7.5–8.5

Table 3.1: Brass electroplating solutions.

thin, decorative layers or for thicker, restorative layers. Applying a thin decorative plate would take one to two minutes and will give a very shiny brass finish. Once it has been rinsed and dried the new brass surface should be lacquered straight away. For a heavy brass plate, times over one hour should be used. When plating for this length of time, an oxide layer will start to form on the surface of the item. After rinsing, this layer is easy to remove by buffing with a soft cloth to reveal a matt brass finish.

BRONZE

Similar to brass plating, bronze plating is most widely used as a protective and decorative strike layer due to its gold-like appearance and increased corrosion resistance. It can vary in colour

3.5a: There are three main types of bronze finish. The first is pink.

3.5b: The second bronze finish is gold.

3.5c: The third bronze finish is white.

3.6: This leaf has been electroformed – covered in conductive paint, electroplated in copper and then finished with a bronze strike plate. This finish gives a gold-like appearance without the use of the expensive precious metal and its cyanide solution.

Just like brass, bronze can be plated from a simple sodium hydroxide solution. The same rate of sodium hydroxide is used, and the priming process is the same, although bronze anodes are used instead of brass.

CHROMIUM

Electroplated chromium, which is used almost exclusively as the final surface deposit, is probably the most important plated metal. Its physical properties, corrosion and wear resistance vastly increase the service life of many parts. These parts would have to be replaced or repaired more frequently, or made from more expensive materials, if not plated in chromium.

There are two main applications for chrome plating; decorative and functional. Decorative chrome plating is usually less than 1μm and gives parts a clean, reflective appearance. This decorative plate also offers slightly increased

	Formula	Electrolyte Composition (grams per litre)		
		Pyrophosphate	Sulphate	Simple Sodium Hydroxide
Copper Pyrophosphate	$Cu_3(PO_4)_2$	5–20		
Stannous Pyrophosphate	$Sn_2P_2O_7$	60		
Sodium Pyrophosphate	$Na_4P_2O_7$	360		
Disodium Phosphate	$Na_4H_2P_2O_7$	20		
Copper Sulphate	$CuSO_4$		30	
Stannous Sulphate	$Sn(11)SO_4$		10–30	
Sulphuric Acid	H_2SO_4		120	
Sodium Hydroxide	NaOH			110 + primed solution
Operating Conditions				
Temperature (°C)		30–60	20	20
Agitation, Filtration		Air or mechanical, coarse filter media	Air or mechanical, coarse filter media	Air or mechanical, coarse filter media
Cathode Current Density		50 A m^{-2}	50 A m^{-2}	50 A m^{-2}
Anodes		Inert	Copper	Bronze (the area of bronze anode to cathode is 1:8)
pH		7.5–8.5	7.5–8.5	7.5–8.5

Table 3.2: Bronze electroplating solutions.

corrosion resistance, lubricity and durability than electroplated nickel, which is the most common material chrome is plated over for this purpose.

Functional chrome surfaces have a thickness greater than decorative plates, thicker than 1μm. This type of coating is used for industrial applications that harness the properties of chrome: heat resistance, hardness, wear resistance, corrosion resistance, erosion resistance and lubricity. In most cases, the appearance of function chrome plates must be of the same quality as the decorative plate.

The oldest type of chrome plating process employs hexavalent chromium, Cr^{6+}, in an aqueous solution. This was followed by solutions containing trivalent chromium, Cr^{3+}, which at the beginning were used mainly for decorative purposes as the thicknesses for functional applications could not be reached. Modern developments of trivalent chrome plating have allowed for thicker layers to be deposited however, Cr^{6+} out performs Cr^{3+} due to the structure of the deposited crystals.

Unlike other metals, chromium cannot be electroplated from a solution that contains only chromium ions – another type of ion is needed for the process to properly work. These ions are catalysts for the electroplating process and are called acid and complexers. For hexavalent chrome, the most common catalysts are sulphate and fluoride. The amount of these catalysing agents is minimal and, as a result their concentrations are relatively difficult to main-

tain. In the case of sulphate, the ratio (by weight) of chromic acid to catalyst acid is 100:1.

The throwing power of hexavalent chromium, when compared to other metals, is low. Trivalent, however, is better. Its throwing power is comparable to that of Watts nickel solution. Despite the lack of throwing power, even and relatively uniform deposits of hexavalent chrome can be achieved using auxiliary anodes, masks and shields.

Chromium Health and Safety

Chromium metal, and most of its oxidations states, are non-toxic, however, solutions containing its ionic form are toxic and very hazardous. Chromic acid is highly irritating and is corrosive to much of the respiratory system and especially damaging to the mucous membranes of the throat and nose. The fumes given in the operating process of hexavalent chromium electroplating must be removed immediately to protect the health of anyone in the area. It can also be carcinogenic and can cause ulcers and dermatitis when in contact with skin. The fumes created by the electrolysis process contain not only oxygen and hydrogen but also all of the constituents of the electrolyte, meaning that if they are not completely removed from the vapour and mist produced, they will escape into the wider environment. The use of wetting agents and fume control balls – small plastic beads – over the surface of the electrolyte will help to reduce the amount of chromic acid given off but are only one part of dealing with the problem. The

3.7: One of the uses of plated copper is shown here, the through hole connections are copper-covered.

3.8: This custom 3D-printed stamp plate has been electroformed in copper, a common electroforming material due to its great versatility and low cost.

3.9: Part-restored antique axe head. The thick copper plate has sufficiently filled pits left by corrosion and it is ready to be sanded and buffed for a smooth finish.

disposal of chromic acid also produces a risk to the environment. These factors lead us to advise strongly against trying this at home. If there are parts that need chrome plating, industrial electroplating plants are much better suited to reduce the risks associated with chromium electroplating.

Copper

Copper is one of the most commonly electroplated metals, after nickel. This is due to the numerous applications in which copper plays a vital role as well as its mechanical, thermal and electrical properties and the ease of electrodeposition. A few of the benefits are listed below:

◆ High plating efficiency + high throwing power: Copper has a high plating efficiency and high throwing power, resulting in excellent coverage on difficult to plate parts such as intricate zinc die castings and difficult geometric shapes. This means that minor imperfections in the base metal, such as pits and scratches, will also be covered.

◆ Inert in other plating solutions: Copper is relatively inert in most plating solutions.

◆ Conductivity: Copper has a very high electrical and thermal conductivity, one of the reasons for its high plating efficiency. The thermal conductivity leads copper to act as an effective thermal expansion barrier. It does this by absorbing and distributing stress when metals with two different thermal expansion rates undergo a change in temperature, resulting in fewer breakages. This is especially useful on plastics.

◆ Range of mechanical hardness: Copper can range in hardness from properties superior to wrought copper to annealed pure copper. It can be relatively soft, which means that the levelling and brightness of copper can be enhanced after plating due to the ease and relative low cost of polishing and buffing.

The qualities above make copper an excellent metal for an underplate. The main applications for copper electrodeposition in larger industries are plating on plastics or other non-metals (electroforming), PCBs, die castings and automotive panels. For home platers, copper is an ideal underplate for restoration work and electroforming.

There are four main types of solution for copper plating: acid, cyanide, alkaline and pyrophosphate. Each type has its own benefits and applications; however, acid copper plating solutions are the widest used because of their relative ease of use, low cost, safety, environmental impact and disposal.

3.10: Copper sulphate crystals, the main constituent of acid sulphate copper electrolyte.

Acid Copper

Acid copper plating has been recorded since the early 1800s, however, it was the invention and expansion of the Daniell cell (mentioned in Chapter 1) that brought copper plating to the interest of scientists. The main driving force behind the development of acid copper solutions was the production of copper-plated electrotypes for printing. This is due to the same properties mentioned earlier, especially the ability of copper to easily plate text and images that are either engraved or in relief.

For home platers, sulphuric acid copper solutions are the most cost-effective choice in terms of power usage and electrolyte formulation compared to the final finish of the copper deposits. Due to the excellent micro-throwing power of acid copper, pits, pores and crevices in all types of surfaces are well filled, improving resistance to corrosion or blistering.

Acid Copper Electrolyte

There are two main types of acid copper plating solution – sulphate or fluoborate. Both electrolytes contain highly ionized copper ions with a valency of two, meaning the ion has the form Cu^{2+}. These ions come from copper sulphate ($CuSO_4$) and copper fluoborate ($Cu[BF_4]_2$) salts. Acid is added into the electrolyte, becoming the second main component after the copper salts. The acids used are sulphuric and fluoboric, with the ions from both aiding the plating system by improving the copper deposits size, increasing conductivity, decreasing anode and cathode polarization and preventing the formation of basic salts on the anodes.

Often, sulphuric solutions are used over fluoborate due to the cost. Sulphuric acid is much cheaper. The concentration of copper sulphate in the electrolyte is not overly critical. High concentrations allow for higher current densities to be used and may aid in grain refinement, that is, produce a fine-grained deposit that has better aesthetical appearance. Lower concentrations decrease the efficiency of the plating system. Overall, these effects

	Formula	Electrolyte Composition (grams per litre)	
		Simple Acid Copper Sulphate	High Throw Sulphate Solutions
Copper Sulphate	$CuSO_4$	200–250	60–100
Sulphuric Acid	H_2SO_4	45–90	180–270
Sodium Chloride	$NaCl$	0.03–0.12	0.05–0.1
Operating Conditions			
Temperature (°C)		18–60 (Optimal 32–43) (Bright Copper below 30)	
Agitation, Filtration		Air or mechanical agitation, batch filtration acceptable, continuous filtration preferred, coarse to fine filter media, 5µm anodes bags suggested	
Additives		Brighteners, Levellers, Carriers	
Cathode Current Density (A dm^{-2})		3.7–5.4	
Anodes		High purity, oxygen-free, optimal phosphorus content 0.02–0.04%	
pH		1	

Table 3.3: Acid copper electroplating solutions.

3.11: Acid copper, plated without brighteners, levellers and other additives, looks dull and pink.

are minimal, however, the concentration of sulphuric acid has a larger impact on the solution and subsequent deposits of copper, becoming increasingly coarse and nodular as the concentration increases.

Chloride ions are added to the acid copper solution. They aid with many properties of the plated copper, including appearance, structure and hardness. They can also help increase the speed of deposition of copper. The range of concentration of chloride ions is strict, between 30–120 parts per million. Above and below this range, the copper plate will begin to become dull and coarse. Increasing the concentration will ultimately stop plating entirely as the anodes will become polarized.

Finally, additional materials can be added to the acid copper electrolyte, such as brighteners, levellers and carriers. These work in the process described in Chapter 1. For the plating of bright, decorative copper, additives must be used. The results are shown in Images 3.11 and 3.12.

Acid Copper Operating Conditions

Temperature: These can vary, anywhere from 18–60°C. However, a temperature between 32 and 40°C is common as it can be maintained economically and with little heating and cooling. Higher temperatures can increase conductivity of the solution. Temperatures below 30°C can also be used to plate bright copper in acid solutions. This temperature is used to maintain good levelling power as some of the components are organic and will denature at temps higher than 30°C.

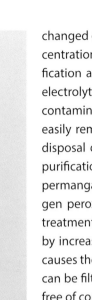

3.12: When the appropriate levels of brighteners are added, copper looks lustrous.

Agitation: This, when balanced with current density, helps with the formation of smaller grain sizes with more desirable properties. Agitation also aids in depositing coatings with uniform thickness and helps with levelling power. There are a few ways agitation can be achieved. Ultrasonic agitation can be used and will help to develop a hard surface layer; however, it is impractical. Another form of agitation is mechanical agitation, that is, physical mixing or moving the electrolyte. Paddles can be used to create a laminar flow of electrolyte to and from the cathode, increasing the uniformity of the copper deposits. Air can also be used to agitate the electrolyte. Air is usually produced by a pump and is coupled with a filtration and purification system. Further information on agitation is given in Chapter 4.

Filtration and Purification: Filtration requirements can change and are usually ruled by the amount of plating that is being done in the tank. Occasional electrolyte filtration can still achieve bright and even deposits on the base material, although continuous filtration is more effective over time and will increase the life of the electrolyte. Filtration depends on the amount of soils introduced into the electrolyte. They can enter through the air used in agitation, brought in by the work or by sludge created by the anode. If the part to be plated is not cleaned thoroughly, it may contain soils on the surface that can dissolve into the solution or create surface films. Anode sludge can be controlled, to reduce the need for filtration, using woven polypropylene anodes bags. These bags collect fine to larger particles of anode sludge and prevent them reaching most of the solution. These bags should be cleaned and

changed often to keep the copper concentration of the anode film high. Purification agents can be added into the electrolyte to react with and oxidize contaminants to particles that are more easily removed. This is essential in the disposal of electrolyte. A few of these purification agents include: potassium permanganate, manganese and hydrogen peroxide. The disposal and waste treatment of acid copper is achieved by increasing the pH to around 9. This causes the copper to precipitate, which can be filtered out leaving the solution free of controlled materials.

Anodes: Rolled and cast copper bars can be used as anodes in copper plating as well as electrolytic copper sheets. High purity oxygen-free anodes are commercially available in various shapes and sizes. Oxygen-free anodes are beneficial as they reduce the amount of sludge produced during the plating process. These types of anodes degrade quickly as the copper is removed easily. Having copper that contains between 0.002–0.004% phosphorous decreases the ease of copper removal while maintaining a high purity, leading to longer lasting anodes. Phosphorized copper anodes are recommended by vendors of copper brighteners when plating bright copper in acid copper solutions. Phosphor-containing copper anodes can occasionally form a black film. This is a matrix of copper ions, chlorine and phosphorus that is filled in with copper electrolyte. This can be removed mechanically with cloth, wire wool or sand paper, but do not need to be if contained within an anode bag. Agitation is critical for the anodes as it promotes the detachment of copper ions

3.13: High-purity copper sheet, cut into an anode.

by replenishing the free acid surrounding the anode. Copper is then dissolved into the solution, maintaining copper levels in the electrolyte, brightness and uniformity of the copper deposits. Common impurities in copper anodes are silver, sulphur, lead, tin and nickel.

Impurities: In comparison with the other plating solutions mentioned in this book, acid copper is more tolerant of impurities. Most metallic ions and particles, contained in the solution due to contamination of previous parts or from anodic impurities, settle to form a sludge at the bottom of the tank. This is because the conditions for copper codeposition are not met. Increased levels of nickel and iron in the electrolyte reduce conductivity of the solution. Arsenic and antimony often codeposit with copper when the correct concentrations are achieved. These cause the resultant plate to become rough

and brittle. Adding gelatine or tannin inhibits the codeposition. Contaminants of tin can cause copper deposits to become smooth. This effect is also seen by some alkali metals and salts. Nitrates in the solution are reduced to ammonia at the cathode, causing roughness on the surface. Oxidizing agents, as mentioned previously, can be added to counteract these. Organic impurities can lead to embrittlement. Organic impurities often come from the decomposition of addition agents such as levellers and brighteners but can be removed using activated carbon followed by intense filtration.

Non-Cyanide Alkaline Copper

The environmental issues surrounding cyanide copper and the cost of disposal led to the development of non-cyanide alkali copper as a replacement. It is

much safer to use, and the waste is much easier to treat. The throwing power of non-cyanide alkali copper is similar to that of cyanide copper, meaning it is almost as good at filling imperfections in the base material. Another benefit of this type of solution is the valency of the copper atom, it has a charge of +2. This charge makes the plating process faster than that of cyanide at the same current and means that higher currents can be used to plate even faster. A downside to non-cyanide alkali copper is that cleaning is critical in the subsequent surface deposition. Adhesion to the base material can be good but a strike plate of pyrophosphate copper is needed first.

Pyrophosphate Copper

Pyrophosphate copper solutions are used in the production of electroformed objects, coating steel, aluminium, zinc die castings and sometimes as a replacement for cyanide copper. The use of pyrophosphate copper in electroforming is still minimal due to the relative expense of the solution as compared to high-throw acid copper solutions.

The solution contains copper pyrophosphate $(Cu_2P_2O_7)$, either potassium or sodium pyrophosphate $(K_4P_2O_7, Na_4P_2O_7)$, nitrate (NO_3) and ammonia (NH_3) along with other additives such as levellers and brighteners. This solution is alkaline, with the copper and salts reacting and forming ions of positive copper and negative pyrophosphate and nitrate.

The ratio of copper to pyrophosphate is critical in the solution, around 3:20. This concentration increases electrical conductivity and promotes anode corrosion. Often, potassium pyrophosphate is used over the sodium salt as it is more soluble and has increased mobility and therefore increased conductivity. The nitrate in the solution also has the same effect of increasing conductivity. Ammonia is added to the solution to give a more lustrous and uniform finish.

The pyrophosphate solution is much more difficult to maintain. The ammonia evaporates from the plating solution, meaning that it must be replenished almost daily. Maintaining the pH and concentration levels can also prove difficult. The high temperature needed between 50–60°C, can be much more difficult to maintain. Overall, acid copper provides the most effective solution for both bright copper plating and high-throw applications such as electroforming.

3.14: Miniature eggs, gold-plated in a mini tank system using a cyanide gold electrolyte. (Lexi Dick, Lexi Dick Jeweller)

Table 3.4: Cyanide gold electroplating solutions.

	Formula	Electrolyte Composition (grams per litre)		
		Alkaline Cyanide Gold	Acid Cyanide Gold	Neutral Cyanide Gold
Gold Potassium Cyanide	$KAu(CN)_2$	6–12	20	20
Potassium Cyanide	KCN	20–30		
Dipotassium Phosphate	K_2HPO_4	20–30		
Potassium Carbonate	K_2CO_3	20–30		
Potassium Hydrogen Citrate	$K_2H(C_6H_5O_7)$		50	
Potassium Hydrogen Phosphate	K_2HPO_4			40
Monopotassium Phosphate	KH_2PO_4			10
Operating Conditions				
Temperature (°C)		50–65	60–70	60–70
Agitation, filtration		Minimal agitation, batch filtration through carbon acceptable, fine filter media,		
Cathode Current Density (A dm^{-2})		1–5	1–6	1–6
Anodes		Gold, Stainless Steel, Carbon, Titanium	Titanium, gold plated titanium (stainless steel, carbon)	
pH		8.5–11.5	4.5–8	6–8

GOLD

As mentioned in Chapter 1, the first written instance of gold electroplating was by Brugnatelli, who electroplated two silver medals with a layer of gold. This decorative application dominated the gold plating industry until the development of the electronics industry, specifically, the manufacturing of special-purpose electrical connections that required gold. The properties of gold at room temperature – excellent thermal and electrical conductivity, ductility, inertness and wear resistance – make it an ideal material for electronic and microelectronic connectors.

Gold electrolyte solutions can be split into two categories; cyanide and non-cyanide. The cyanide baths can be further split into alkaline, acid and neutral. These solutions can form gold deposits that are hard or soft, dull or bright. The main downside to using cyanide-based solutions is obviously the formation of cyanide complexes, especially cyanide gas, which is highly toxic. We have avoided cyanide solutions so far because of their impact, however, cyanide-free gold solutions are difficult to control and use in a home environment and thus we will now discuss the use of cyanide-containing solutions. If using these make sure you read all the required safety data sheets (MSDS) and take all necessary safety precautions. For further information see Chapter 9. It may be easier to purchase pre-made gold plating solutions than to make them from scratch.

Alkaline Cyanide Gold

Alkaline cyanide electrolytes have a high pH, usually greater than 8.5. In the higher pH range, the potassium cyanide in the bath breaks down to cyanide ions, also called free cyanide. When there is free cyanide, gold can be used as a sacrificial anode. This is because gold can be electrochemically oxidized to form gold cyanide complexes. When the pH drops below 8 there is no free cyanide and the gold anode would act like an inert anode. Another effect of a pH below 8 is that hydrogen cyanide gas evolves from the free cyanide in the electrolyte, which is incredibly hazardous.

Acid Cyanide Gold

First developed for the jewellery industry, acid cyanide plating baths operate at a pH of around 4.5. Unlike the alkaline cyanide above, this solution is capable of plating polymers and organic materials without attacking them. Acid cyanide is therefore widely used in the electronics industry for contact surfaces, corrosion protections and bonding surfaces.

As the pH is low, anodes for this type of gold electrolyte are usually titanium or gold-plated titanium. Stainless steel and carbon anodes can be used but often contaminate the solution due to the anodic reaction of the solution. As the anodes are insoluble, more gold potassium cyanide must be added on a regular basis to maintain the gold content and to maintain the pH levels. Citric acid or potassium hydroxide can also be used to maintain pH levels.

Acid gold is the best solution for codeposition with other metals to create gold alloys. For specific applications, certain properties of gold must be achieved. Hard and bright deposits can be achieved by adding nickel or cobalt salts into the solution; these metals act as brighteners. The concentrations needed are very low, only around 0.07–0.1 and 1g per litre for cobalt and nickel respectively. Iron can also be introduced into the electrolyte to harden the electroplated gold, however, the newly plated surface becomes much more brittle.

3.15: Iron-plated electrotype, used as a long-lasting stamp.

Neutral Cyanide Gold

Overall, the best cyanide-containing gold solution for altering the physical and chemical properties of the newly plated surface is a neutral solution. Hard gold can be achieved through the manipulation of the operating conditions rather than the addition of other metals. These solutions must be checked regularly as control of temperature, gold concentration and pH level is critical in the formation of hard gold. Neutral cyanide gold contains the same amount of potassium gold cyanide as acid gold, but it has potassium hydrogen phosphate and monopotassium phosphate. The pH range means there is no free cyanide in the solution, making it the best for plating electronics with no etching of organic polymers.

Non-Cyanide Gold

There are several shortcomings to using cyanide in gold, and one major reason other methods have been developed; the use of cyanide. The health and safety of workers and the environmental impact of working with cyanide is a major concern.

The main constituent of non-cyanide baths is gold sulphite. This type of solution allows for the plating of low-stress, soft and bright deposits. One downside to using gold sulphite is the possibility of spontaneous decomposition of the electrolyte; gold particles form and drop out of the solution. Additives can be used to increase the stability of the solution.

IRON

Electroplating of iron was used at the turn of the twentieth century for numerous applications: electrotypes, ammunition production, iron electroforming and restoration.

The main driving force for its use and development are the properties of electroplated iron and for its low cost. It was often used in place of copper and nickel in the Second World War to produce various pieces of typing equipment, leading to the conservation of the more expensive metals. Even though many of the applications iron electroplating has been used for were short-lived, interest persists. It is still an important metal to study, especially within electroplating, due to its abundance, low cost and massive range of physical properties that can be produced, as well as its magnetic properties. There are a few downsides to iron electroplating. It has very limited usage in terms of specialized or high-volume applications. While iron itself is inexpensive and an

Table 3.5: Iron electroplating solutions.

	Formula	Electrolyte Composition (grams per litre)		
		Iron Ammonium Double Salt Sulphate	Sulphate	Chloride
Iron Ammonium Sulphate	$FeSO_4 (NH_4) SO_4$	250–300		
Iron Sulphate	$FeSO_4$		250	
Ammonium Sulphate	$(NH_4)SO_4$		120	
Iron Chloride	$FeCl_2$			250–450
Calcium Chloride	$CaCl_2$			150–190
Operating Conditions				
Temperature (°C)		25	60	88–90
Agitation, filtration		Minimal agitation, batch filtration through carbon acceptable, fine filter media		
Cathode Current Density (A dm^{-2})		2	4–10	2–9
Anodes		Iron – anodes must be removed due to further dissolution in electrolyte		
pH		2.8–3.4	2.1–2.4	0.2–1.8
		4.0–5.5		

3.16: For a simple iron plating solution all that is needed is ammonium chloride and an iron anode. The solution should be primed in the same way as the simple brass solution from the beginning of this chapter.

iron electrolyte is also inexpensive, the cost to maintain an iron plating system is high, negating the low material cost.

Practically all iron electrolytes are acidic. The solutions contain iron salts in the ferrous state, that is *Fe(11)*, with either sulphate or chloride or a mixture of the two. Fluoborate and sulphamate electrolytes have also been used successfully. Table 3.5 gives a table of common iron electroplating solutions.

Iron sulphate plating systems

produce smooth and light grey deposits. They have the advantage of being able to plate at room temperatures, have very little pitting and can build up thick deposits. While the pH range of the sulphate bath is quite large, at around 3.5 iron hydroxide starts to precipitate, leading to dark-coloured and stressed deposits. In the rest of the pH range, the deposits can be brittle, and the deposition rate is very slow compared with a hot chloride electrolyte.

Iron chloride baths, as mentioned, have a high deposition rate. This is due to the higher current density possible in the solution when used at higher temperatures – above 85°C. The most common iron chloride bath is a mix of iron and calcium chloride and is referred to as the Fischer-Langbein solution. The contents of the solution are in the table above. The addition of calcium chloride increases conductivity of the solution and has a very small effect on the structure of the deposited iron crystals. When used at lower temperatures, around 25°C, this iron electrolyte produces deposits that are dark, hard and highly internally stressed. As the temperature increases, the colour becomes lighter.

There are various other acid iron electrolytes, such as fluoborate, sulphamate and sulphate-chloride, as well as alkaline baths, but these are for more specialized purposes, mainly for the production of electrotypes and strip materials. There are various metals and materials that can be codeposited with iron to form alloys, including cobalt, nickel, platinum, vanadium, zinc, carbon, phosphorus and boron. The most common alloy is that of iron

and nickel. It is used as a decorative and protective undercoat to a thin plate of chromium. It is chosen for its operating cost, which is lower than nickel alone. These baths usually consist of nickel chloride, nickel sulphate, ferrous sulphate, boric acid, brighteners and stabilizers.

NICKEL

Nickel plating began to be developed in the late 1830s and early 1840s, with the first practical electroplating solution being formulated by Bottger. His formulation laid the foundations for commercial use. Nickel plating solutions were further pushed forward by Dr Isaac Adams. His formulation, similar to Bottger's, had an emphasis on the pH of the electrolyte. This was vital for controlling the quality of the

nickel deposits. The most substantial development within nickel plating was made by Professor Oliver P. Watts, who formulated an electrolyte made from nickel sulphate, nickel chloride and boric acid. His formulation led to the elimination of previous formulations due to its numerous advantages.

Nickel electroplating is similar in set-up to other processes that employ soluble metal anodes; direct current travels between two electrodes immersed in a conductive electrolyte made from nickel salts.

Nickel is the most commonly used metal for electroplating due to its versatility and engineering characteristics. The applications fall into three main categories: decorative, functional, electroforming.

Table 3.6: Decorative nickel electroplating solutions.

	Formula	Electrolyte Composition (grams per litre)		
		Watts Nickel	Nickel Sulphamate	Semibright Nickel (bright with the addition of additives)
Nickel Sulphate	$NiSO_4$	225–400		300
Nickel Chloride	$NiCl_2$	30–60	300–450	35
Nickel Sulphamate	$Ni(SO_2NH_2)_2$ or $H_4N_2NiO_6S_2$		30–45	
Boric Acid		30–45	0–30	45
Operating Conditions				
Temperature (°C)		44–66	32–60	54
Agitation, filtration		Air or mechanical agitation, continuous filtrations needed, fine filter media, carbon filter		
Cathode Current Density		3–11	0.5–30	3–10
Anodes		Nickel		
pH		2–4.5	3.5–5	3.5–4.5

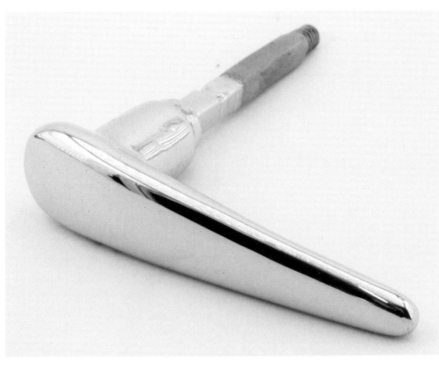

3.17: This handle shows how mirror-finished a bright nickel layer can be.

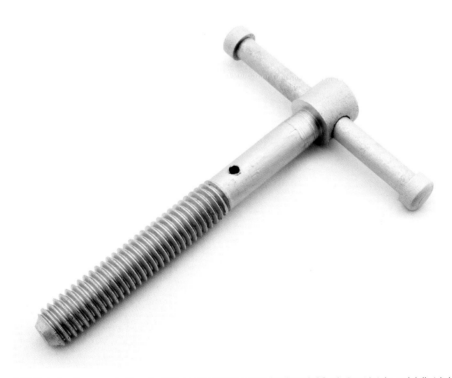

3.18: Reducing the concentration of additives will produce a semi-bright nickel finish. Semi-bright and dull nickel finishes are usually used on items that are more functional than decorative.

Decorative Nickel Plating

Modern decorative nickel electrolytes vary little from the original Watts formulation, with only the addition of organic (and a few metallic) compounds used as levellers, carriers and brighteners. The only competition to the Watts solution is a nickel sulphamate solution. Both solutions can be used as functional and electroforming electrolytes, however, the Watts solution is still the basis for decorative nickel plating. The table below shows the typical nickel solutions and their contents, as well as operating conditions.

There are different types of decorative nickel solutions available to electroplaters, each with great corrosion resistance and lustrous finish. The easiest to make, use and maintain is a Watts-based bright nickel solution.

Modern bright nickel solutions are based on the original Watts solution, except they contain different additives to produce bright deposits. These solutions have excellent macro and micro-levelling capabilities, giving a uniform and bright surface to badly scratched and pitted parts. These solutions also form deposits with fair ductility and with few internal stresses. The additives for a bright nickel electrolyte fall into three categories: carriers, brighteners and auxiliary brighteners.

Carriers are usually aromatic organic compounds that are a source of sulphur, which is codeposited with nickel. Sulphur ions are much larger than nickel ions and so, when deposited into the crystalline lattice structure, act to change the natural internal stress of the nickel from tensile (contractive stress where the atoms are pulling towards each other) to compressive (expansive stress where the atoms push away from each other). Carriers also act to refine grain size and provide coatings with increased lustre. Carriers will not form a bright surface layer on their own. They are not used up in the electroplating process but are removed through dragout or when the electrolyte is filtered through activated carbon.

Brighteners, when used in conjunction with carriers and auxiliary brighteners, give the electrolyte great levelling qualities and bright deposits. There are a vast number of different brighteners available but they all tend to increase internal stress and brittleness of the deposits and so the concentration is kept low.

Auxiliary brighteners increase the effects of carriers and brighteners, giving increased lustre, brightening and levelling. They must be used in lower concentrations.

Semibright nickel electrolyte solutions are all based on the Watts nickel solution. A typical make-up is shown in the table above, the only addition being levelling agents. A plate with this solution gives smooth deposits that are semi-lustrous. This highly levelled surface is easy to polish to a mirror-bright finish. The main benefits of semi-bright nickel are its corrosion resistance and ductility, making it critical in situations where the part comes under stress. It is also the basis for multi-layer nickel.

	Composition		pH	Temperature (°C)	Cathode Current Density (A dm⁻²)
	Material	Quantity (g L⁻¹)			
All Chloride	Nickel Chloride	225–300	1–4	50–70	2.5–10
	Boric Acid	30–35			
All Sulphate	Nickel Sulphate	225–410	1.5–4	38=70	1–10
	Boric Acid	30–45			
Black Nickel (chloride)	Nickel Chloride	75	5	24–32	0.15–0.6
	Zinc Chloride	30			
	Ammonium Chloride	30			
	Sodium Thiocyanate	15			
Black Nickel (sulphate)	Nickel Sulphate	75	5.6	24–32	0.15
	Zinc Sulphate	30			
	Ammonium Sulphate	35			
	Sodium Thiocyanate	15			
Fluoborate	Nickel Fluoborate	225–300	2.5–4	38–70	3–30
	Nickel Chloride	0–15			
	Boric Acid	15–30			
Hard Nickel	Nickel Sulphate	180	5.6–5.9	43–60	2–10
	Ammonium Chloride	25			
	Boric Acid	30			
High Sulphate	Nickel Sulphate	75–110	5.3–5.8	20–32	0.5–2.5
	Sodium Sulphate	75–110			
	Ammonium Chloride	15–35			
	Boric Acid	15			
Phosphorus Nickel	Nickel Sulphate	170 or 330	0.5–3	60–95	2–5
	Nickel Chloridie	35–55			
	Boric Acid	0 or 4			
	Phosphoric Acid	50 or 0			
	Phosphorus Acid	2–40			
Sulphate/ Chloride	Nickel Sulphate	150–225	1.5–2.5	43–52	2.5–15
	Nickel Chloride	150–225			
	Boric Acid	30–45			

Table 3.7: Nickel electroplating solutions.

Other Nickel Solution

Watts nickel solutions are not the only available solutions to be used at home. There are numerous other electrolytes and each one has been formulated to meet specific requirements.

All Chloride

The main advantage of an all-chloride nickel solution is the ability to use it with high cathode current densities, reducing the anode material required in tank set-up. It also has a high conductivity and has a slightly better throwing power over Watts nickel. The physical properties given by an all-chloride solution are also better than Watts nickel; the grain size and final surface finish is finer and smoother, harder and stronger. One of the main disadvantages of this solution is the corrosive fumes given off. Without proper protection, the mist will attack any equipment it encounters, such as vents and extraction fans.

All Sulphate

All sulphate solutions are designed mainly to be used with insoluble anodes. As an example, insoluble auxiliary anodes may be used to plate low current density areas such as the inside of light reflectors, pipes or tubing. The anodes can be made from a few materials; lead, carbon, graphite and platinum. As nickel is deposited, oxygen is created at the anode, decreasing both pH and nickel concentration. Nickel carbonate is added to maintain both properties.

Black Nickel

The two types of black nickel solution were designed for decorative purposes. They offer little in the way of corrosion or abrasion resistance and are usually deposited over a bright or semibright nickel surface.

3.19: Example of a semi-bright black nickel finish. Black nickel can become lustrous when applied on the top of a bright nickel surface.

Fluoborate

The nickel fluoborate solution is similar to the Watts nickel. It can be used over a wide range of concentrations, temperatures and pH, meaning it is relatively simple to use. Fluoborate has the advantage of being able to be used at very high current densities, leading to quicker plating. The main disadvantage is that the solution can be chemically corrosive; the fluoborate ion will attack some materials.

Hard Nickel

The hard nickel solution was developed specifically for its mechanical properties and functional applications. The outcomes of the solution are controlled hardness, improved abrasion resistance, higher tensile strength, good ductility. The solution must be monitored and maintained to give reproducible results. The nickel deposits can also become nodular.

High Sulphate

High sulphate nickel solutions have been developed primarily for the plating of zinc die-cast pieces. The solution does very little corrosive damage to the nickel compared with Watts nickel due to the high-sulphate and low-nickel ion concentration and low pH. This mix also creates good throwing power. The layer of nickel that is deposited is highly stressed and therefore is usually applied to a layer of copper that can absorb the stress created by the newly formed layer.

Phosphorus

The nickel phosphorus solution codeposits nickel with phosphorus, creating an alloy that has enhanced corrosion resistance. The solution produces a very similar result to electroless plating of nickel.

Sulphate/Chloride

The sulphate/chloride solution was designed to overcome some of the shortcomings of the all-chloride solution. It contains equal amounts of nickel sulphate and nickel chloride. The properties of the sulphate/chloride solution lie in the middle between Watts and all chloride, except it has a higher conductivity, can be used at high current densities and has a higher internal stress. This stress is still lower than an all-chloride solution.

Electroplated Nickel Applications

Nickel is most often used for decorative applications. Currently, around 80% of all nickel used in electroplating is used to decorate parts. These parts can be for a range of applications, most notably automotive parts, building hardware, tools, housewares, furniture, bicycles, motorbikes and mopeds. The other 20% is used in electroforming and for functional parts.

The main reason for using nickel for a functional part is to improve certain qualities, most usually corrosion resistance, hardness, wear resistance and magnetic properties. While a defect-free appearance is still important, the lustrous, mirror-like shine is not. The two main functional nickel electroplating solutions used are the Watts nickel and sulphamate nickel. These are included in the table above.

Nickel Anode Material

Almost all nickel plating systems use soluble nickel anodes. Nickel, made specifically for anode material, does not usually introduce impurities into the electrolyte, they only work to add ions and maintain current. Most nickel anode material contains above 99% nickel with around 0.5% nickel oxide and traces of sulphur. Sulphur is added into the anode material to aid in dissolution. Sulphur lowers the voltage needed to dissolve nickel ions and helps to create a more uniform dissolution. When sulphur-containing nickel anodes are dissolved, an anode film is created that in turn creates a sludge. However, this sludge is only 0.1% of the total anode material dissolved. The lower voltage needs of sulphur-containing nickel help to reduce cost by conserving energy as compared to other types of nickel.

One problem with nickel anodes is the rate and non-uniformity of dissolution. This means that the area and anode potential change over time, which causes problems at the cathode. As an effort to reduce this effect, titanium anode baskets were introduced. These baskets, made from a skeleton of titanium wire and coated in anode bag material, improve plating quality by ensuring the area of the anode changes very little. The material on the

3.20: Typical nickel anode material, pure squares of nickel that have been drilled and hung using titanium wire. To achieve the necessary anode surface area, more of these small squares must be placed in the tank.

outside of the basket facilitates the free flow of electrolyte solution, nickel ions and other ions while also preventing larger pieces of nickel from falling into the solution. For large-scale applications, titanium baskets are now critical in maintaining current distribution, ion transfer and ultimately workflow as the replenishment of anode material is simple and does not involve removal of the anodes and disruption of work. Titanium anode baskets have the added benefit of being able to take any shape, altering the current density on the cathode as needed. Titanium can corrode in nickel solutions, especially if not in contact with nickel. Nickel protects the titanium. Titanium cannot be used in fluoborate nickel solutions or solutions containing fluoride ions as they activate the titanium, causing it to corrode.

Table 3.8: Cyanide silver electroplating solutions.

	Formula	Electrolyte Composition (grams per litre)	
		Bright (low silver concentration)	High Speed (high silver concentration)
Silver Potassium Cyanide	$KAg(CN)_2$	31–55	45–150
Potassium Cyanide	KCN	50–80	70–230
Potassium Carbonate	K_2CO_3	15–19	15–19
Potassium Nitrate	KNO_3		40–60
Potassium Hydroxide	KOH		50
Operating Conditions			
Temperature (°C)		20–50	
Agitation, filtration		Minimal agitation, batch filtration through carbon acceptable, fine filter media	
Cathode Current Density (A dm^{-2})		1–2	1–6
Anodes		99.98% Silver, Stainless Steel, Steel, Carbon, Titanium	
pH		11	

Nickel Plating Bath Control

One of the most important factors in nickel plating and the quality of the nickel deposited is the control of the plating bath. As shown in the table above, the range of some constituents or properties of nickel solutions can be very fine or absolute. This control starts before any plating has begun. When the bath is set up, pH must be checked and maintained to achieve the best results. Regular checks and adjustments during the use of the bath should be continued.

The pH of a nickel plating solution will rise due to regular plating activity. This is due to the cathode efficiency being below 100% and the build-up of nickel ions in the solutions. There is therefore a necessity to check regu-larly the pH of a nickel plating bath that undergoes continuous plating. Acid is added to bring the pH back to regular levels. If the opposite is occurring, the pH is decreasing, this is an indication that something is wrong with the plating solution. In a Watts nickel bath, sulphuric acid is added for pH adjustment. In sulphamate, sulphamic acid is added. Boric acid is added to the other nickel plating solutions.

The operating temperature of a nickel solution has a large impact on the surface deposited. In general, most plating baths are maintained between 40 and 60°C. The properties of the newly formed layer will begin to deteriorate if the temperature is outside the range specified. Additionally, addition agents are often organic and will not work above certain temperatures, so it is best to check on the manufacturers' specifications as to the accepted temperature range for their product.

SILVER

Electroplating using silver has been around for a similar amount of time as gold, however, the first patent was made in 1840 by the Elkingtons. The bath in the patent comprised of a silver cyanide complex and is very similar to industrial solutions used today, however, there non-cyanide solutions have also been created. The use of electroplating silver varies and ranges from decorative tableware, where the wear resistance of silver is critical, to electronic components and mirrors. There are two methods to plating silver: with and without electricity.

Electrodeposition of Silver

The type of electrolyte used in silver plating is dependent upon the final application. For decorative purposes, electrolytes with low concentrations of silver ions are preferred. This is compared to engineering requirements, which require thicker deposits and therefore higher concentrations of silver. These two typical silver solutions are shown in the table below. Similar to the gold solutions, these contain cyanide, so make sure to follow all the safety precautions. Both silver solutions operate at lower temperatures, between 20 and 50°C, they are alkaline, and should have a pH of around 11.

The first patented silver electrolyte was cyanide based. Present-day silver solutions have changed very little, still using silver cyanide. However, cyanide-free alternatives are now greatly improving in final surface quality and deposition. These non-cyanide solutions, based on nitrates, thiourea and other similar compounds, are not as stable as they have low adhesion and are, in general, difficult to control and have poor results. Therefore, despite its hazards, cyanide silver is the only solution used commercially.

The finish of a bright silver cyanide plating solution can be seen in Image 3.21. In practice the bright silver solution, with high concentrations of free cyanide, has a higher throwing power, making it ideal for intricate items, such as jewellery pieces. However, the silver layer deposited is only around 2 to 5μm thick. This is because of the need for lower current densities. Higher concentrations of silver allow for the use of

3.21: Silver-plated goose figure, plated using the same method as the golden eggs, a mini tank plating system using a silver cyanide solution. (Lexi Dick, Lexi Dick Jeweller)

higher current densities and therefore, give a quicker overall plating speed.

Additives, such as brighteners and levellers, are often added to silver plating solutions. There is a large range available, but they are mostly based on sulphur-containing organic compounds. These compounds help to produce bright and lustrous surfaces that either reduce or stop the need for buffing and polishing, making the process more efficient in time and waste material.

The anode and cathode efficiencies are close to 100% – meaning the same amount of silver that is dissolved from the anode is deposited on the cathode. Having such a high efficiency means that high purity silver should be used for the anodes; 99.98% is recommended. Having both pure anodes and a high efficiency means that the solution can be kept in balance for a reasonably long time and gives rise to a stable electrolyte solution, making it easy to maintain.

Before Silver Plating

The pre-treatment of metals before silver plating should be followed as per Chapter 6, with the addition of a silver strike prior to plating. A strike is needed as most metals are less noble than silver. The lower nobility means the other metals are more likely to react to the solution, simply by immersion with no current. This causes precipitation of silver; the silver ions will be replaced by the other metal and will form silver particles, which are not dissolved in the solution. Poor adhesion of a silver plate would be the result. Strike plates are therefore needed to achieve acceptable adhesion and a good quality of plate.

The silver strike solutions contain very low concentrations of silver ions and high cyanide ion concentrations. This mix lowers the chance of silver dropping out of the solution. A typical strike solution for non-ferrous metals, such as nickel and copper, works in the same way as the bright silver solutions in the previous table. The concentrations differ; silver potassium cyanide should be 1.5 – 5 gL^{-1} and potassium cyanide should be 75 – 90 gL^{-1}.

	Formula	Electrolyte Composition (grams per litre)		
		Alkali Tin – Potassium Stannate	Alkali Tin – Sodium Stannate	Acid Tin – Sulphate
Potassium Tin Hydroxide	$K_2Sn(OH)_6$	95–110		
Potassium Hydroxide	KOH	15–30		
Sodium Tin Hydroxide	$Na_2Sn(OH)_6$		95–110	
Sodium Hydroxide	NaOH		7.5–11.5	
Tin Sulphate	$SnSO_4$			15–45
Tin	Sn			7.2–22.5
Sulphuric Acid	H_2SO_4			135–210
Operating Conditions				
Temperature (°C)		60–90		18–25
Agitation, filtration		Mechanical agitation, batch filtration through carbon acceptable, continuous filtration best, coarse to fine filter media		
Cathode Current Density (A dm^{-2})		0.32–1	0.05–0.32	0.1–2.5
Anodes		Tin		
pH		8–13		1–3

Table 3.9: Tin electroplating solutions.

are non-corrosive, making them safer to work with than most electroplating solutions. They also do not need other additives, brighteners, levellers and so forth, making tin electrolytes easier to create and control.

There are a few downsides to using this kind of tin solution. It must be run at a higher temperature. In fact, alkali stannate electrolytes give good results at 65°C and above. Below this temperature, two major problems occur that impact plating. First, tin begins to precipitate from the solution, falling to the bottom and reducing the stannate content. Second, the tin anodes become passive, leading to minimal dissolution of tin. In addition to the temperature problem, tin ions in the alkali solution are in their trivalent form, meaning that around twice as much amperage is needed to plate the same quantity of tin over a specific time than that of the acidic solution (which is divalent).

TIN

There are two main types of electroplating solutions for tin: alkali stannate based and acidic stannous salt based.

Alkaline Tin

The first solution, alkaline, contains either sodium or potassium stannate and a corresponding alkali metal hydroxide group (sodium hydroxide or potassium hydroxide). These solutions

Acidic Tin

As just mentioned, electroplating with an acidic tin solution uses less electricity than the alkali solution. This is one of

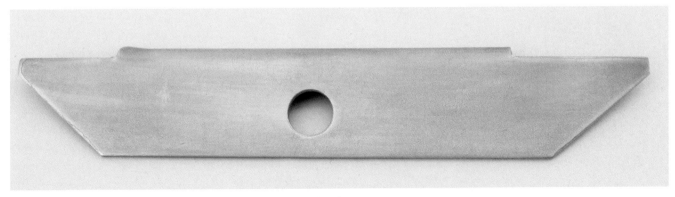

3.22: Tin-plated items look very similar to steel when dull or to bright acid zinc when shiny.

3.23: The main application for zinc plating is to increase the corrosion resistance of fasteners such as these. There bright zinc surfaces show the typical zinc colour with the addition of a clear chrome passivate.

3.24: A zinc-plated mounting bracket that has been left to oxidize in the air, becoming dull and grey.

discrete circular deposits, which only adhere loosely to the surface and do not fully cover the work.

To improve upon these deposits, an array of additives must be used. They are mainly organic and can include gelatine, cresylic acid, cresol sulphonic acid and aromatic hydroxyl compounds [10]. The addition of multiple additives generally leads to the acidic solution being difficult to control.

The most common acidic tin solution is sulphuric acid solution. This is due to its low cost and smaller environmental impact. The additives used in a sulphuric acid bath differ slightly from regular baths and can include tartrate, tartaric acid, formaldehyde and various ethers [11]. Again, these can be difficult to control. For the purposes of home plating it would be easier to use the alkali tin solution.

ZINC

Zinc is one of the top metals used in electroplating. In fact, approximately half the world's consumption of zinc is used for the coating of zinc on steel, with electrodeposition of steel sheet alone accounting for between 25 and 30% of zinc consumption. Most of the remaining zinc is used to plate fastenings and fixtures, steel and iron parts to increase their corrosion resistance. Once plated and without further treatments, zinc plating will become dull grey with exposure to air. The air causes the surface of the zinc to oxidize, leading to a rough texture and dullness. For some applications, this oxide layer may not be too much of a problem, but it will eventually lead to full corrosion

the only benefits of this solution.

The main disadvantage of acidic stannous salt solutions is that they have a critically complicated composi-

tion. Attempting to plate with simple acidic solutions gives poor deposits with either nodules, pits, scorching or delamination. These usually occur as

of the zinc. That is why, for bright zinc especially, a post-plating coating is applied. Most commonly, this coating is a chromate conversion coating or lacquer and is used to add extra corrosion resistance, preserve surface finish or both. The topic of post-plating treatments is covered further in Chapter 8.

Commercial zinc plating electrolytes are varied, with baths comprising of cyanide, alkaline non-cyanide, acid sulphate and acid chloride. Until relatively recently, the 1970s, most zinc plating was done using cyanide baths. There has been a very big international push, however, to remove cyanide from plating baths for environmental and health reasons. This has led to the development of other zinc plating electrolytes. Today, the two most commonly used are alkaline non-cyanide zinc and acid zinc.

Alkaline Non-Cyanide Zinc

Alkaline non-cyanide zinc plating baths are a logical progression from micro-cyanide baths towards creating safer zinc plating baths. In the early development of alkaline zinc electrolytes, the deposited layer was poor, having low levels of adhesion both to the base material and between zinc crystals. The result was a low-density, powdery surface that was of little practical use. Experiments with the supply of electricity (using pulsated currents or superimposition of alternating on direct) and organic additives led to the development of better-quality, dense zinc surfaces.

The composition of alkaline baths is relatively simple, containing only zinc,

Table 3.10: Alkaline non-cyanide zinc electroplating solutions.

	Formula	Electrolyte Composition (grams per litre)		
		Alkaline Low Chemistry	Alkaline High Chemistry	Alkaline Regular
Zinc Metal – dissolved from anodes	Zn	3–9	13.5–22.5	6–17
Sodium Hydroxide	NaOH	80	75	75–150
Additives		Use as specified by suppliers		
Operating Conditions				
Temperature (°C)		15–55		
Agitation, filtration		Air or mechanical agitation, batch filtration through carbon acceptable, continuous filtration best, coarse to fine filter media		
Cathode Current Density (A dm^{-2})		0.35–5		
Anodes		High Purity Zinc		
pH		10		

sodium hydroxide and additives. This is the same system as the simple bronze and brass solutions mentioned earlier in the chapter and a primed solution should be created in the same way. The table above shows the composition of three types of solution: a low-chemistry (LC) solution used for rack and barrel plating, a high-chemistry (HC) solution and a regular solution, used for manual and home plating.

The window of concentrations for the components of an alkaline zinc bath is slim – especially the zinc concentration. For example, in LC baths, a zinc concentration of 7.5 gL^{-1} gives bright surfaces with an efficiency of around 80%. Decreasing this concentration will also decrease the efficiency but keep the surface relatively bright. In contrast, raising the zinc concentration to high levels, around 17 gL^{-1}, will begin to give a dull grey finish. Increasing the level of sodium hydroxide increases efficiency by increasing conductivity

but it also lowers the throwing power, meaning that thicker deposits of zinc build up on edges and corners. While alkaline zinc plating has grown and is very successful as a method of depositing zinc, it is surpassed in terms of usage by acid zinc.

Acid Zinc

Acid zinc has had a massive influence in the industry since its development, and now accounts for around 50% of all industrial rack and barrel plating systems. This is due to the numerous benefits of the electrolyte. The first of these is the ease of neutralization for disposal; the solution can be neutralized easily and is then safer to handle. The current efficiencies are high and, coupled with a much higher conductivity, energy costs are substantially reduced in comparison to alkaline and cyanide solutions. The surface brightness straight out of the bath, due to the

	Formula	Electrolyte Composition (grams per litre)			
		All Ammonium Chloride	Low Ammonium		Potassium Chloride (no ammonium)
			Potassium Chloride	Sodium Chloride	
Zinc metal (Zinc Chloride)	Zn ($ZnCl_2$)	15–30 (45–90)	15–30 (45–90)	15–30 (45–90)	22–38 (66–114)
Ammonium Chloride	NH_4Cl	120–180	30–45	30–45	
Potassium Chloride	KCl		120–150		185–225
Sodium Chloride	NaCl			120	
Boric Acid	H_2BO_2				22–38
Operating Conditions					
Temperature (°C)		15–55			
Agitation, filtration		Air or mechanical agitation, batch filtration through carbon acceptable, continuous filtration best, coarse to fine filter media			
Cathode Current Density (A dm^{-2})		0.3–5			
Anodes		High Purity Zinc - insoluble anodes, stainless steel, lead etc., cannot be used in chloride electrolytes as chlorine gas is evolved at the anode which is very dangerous			
pH		5–6			

Table 3.11: Acid chloride zinc electrolyte solutions.

due to the complexing nature of the ammonium ion, and they require extensive chlorination to make them safe. This is the reason for the development of potassium baths.

There are various other types of zinc plating solution: sulphate, fluoborate, sulphamate, acetate, perchlorate and molten salts. They are not in much use and do not have the beneficial properties of the chloride acid zinc solution.

Zinc Alloy

While pure zinc plating is still commonplace for barrel and rack applications for plating fastenings and other mass-produced parts, zinc alloy plating is widely used in answer to demands for higher quality and longer-lasting finishes.

The most common metal that is codeposited with zinc is nickel. Zinc-nickel alloys can be plated from both acid and alkaline solutions. The amount

presence of organic additives, is excellent, and parts will have lustrous, mirror-like shine. It can electroplate more metals than cyanide or alkaline baths, including iron and carbonized parts. It is also the best zinc solution for home plating due to its cost-effectiveness and ease of operation.

In almost all zinc plating baths, additives are employed to aid with surface finish. In acid zinc, the addition of additives is critical in producing bright and level deposits. These additives fall into the two categories of carriers and brighteners. They are proprietary mixes of organic compounds usually consisting of polyalcohol, polyamine, fatty alcohols, polyglycol ether and quaternary nitrogen bonds as carriers and aliphatic, aromati, and heterocyclic carbonyl-bonded molecules for brighteners. Carriers are essential as they often make the brighteners soluble in the solution as well as reduce the surface energy of the solution.

There are two main disadvantages of acid zinc. The first of these is that the electrolyte is corrosive, meaning that it must be treated carefully, and the fumes extracted well when in use. The second is that the zinc deposits are prone to entrapment of the electrolyte solution, eventually leading to bleed-out, which can ruin a part. As an addition, the disposal of ammonium-containing solutions can be tricky

3.25: Half-size Zenith carburettor for a V8 aero engine replica. Fabricated from brass, the main body was then plated with acid zinc to give an authentic and original-looking finish.

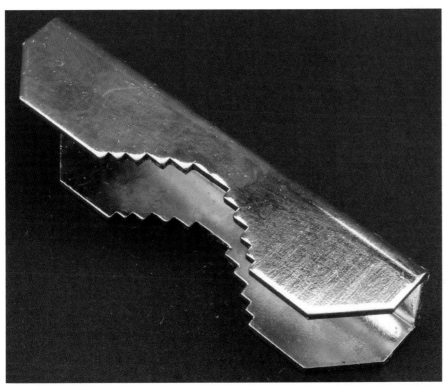

3.26: An example of a zinc-nickel alloy-plated item. The finish is very similar to that of regular zinc with the addition of extra corrosion resistance. This part can be passivated in the same way as a regular zinc surface.

of nickel deposited ranges from 10–14% and 6–9% for acid and alkaline baths respectively. The main downside of acid zinc-nickel plating is that the thickness distribution of the nickel is low. This means that nickel tends to be deposited on higher current density areas, peaks, edges and corners, which leads to variations in the alloy. For most applications, this effect is insignificant, especially when it comes to an aesthetical appearance.

To plate acid chloride zinc nickel, simply add 50g per litre into the solution. For alkaline zinc, add nickel anodes when creating and priming the solution.

There are a few more zinc alloys that can be plated but are not as prevalent as nickel. Cobalt can be added to low-ammonium or ammonium-free zinc electrolyte mixes at a rate of 1% by volume to produce a zinc-cobalt alloy. This gives a good current efficiency and so a high rate of plating. Iron can be added into a zinc electrolyte, usually alkaline. Added at a rate of 0.3–0.9%, iron produces a non-silver, black chromate when a chromate conversion coating is applied. Tin can also be plated with zinc. It is usual for the ratio to be about 70:30 tin to zinc, meaning that it is technically more of a tin alloy. This alloy has great frictional properties and ductility after plating, and is also very resistant to atmospheres containing sulphates. A common zinc alloy that is plated is brass, which has already been covered at the start of this chapter.

4 Electroplating Tank Set-up

Now that you know about the different types of solution, how they are composed and their operating conditions, it's time to think about what you are going to put them in. In this chapter, we will cover the things you should consider when choosing a plating tank. What material should it be made from? What size do you need? Deciding the type of metal you want to deposit and the objects you want to plate will determine the right tank for you. Thinking further than just the tank, what else will you need for electroplating? Well, in almost all situations, having a heater will benefit your plating system; the finer details of why are described later. Filtration and agitation are two important processes that, again, have a major impact on your plating system and the results. There are various methods and pieces of equipment you can use to achieve both filtration and agitation; these will be described in later sections in this chapter. A discussion of power supplies and working out current needs will also be discussed.

4.1: Having a tank that is too small will often lead to problems with current density, meaning that the edges may over-plate and become scorched. Having a large tank with a large anode-cathode separation is ideal in terms of current density. This is because the distance to the anode of peaks and troughs on an item become comparably small as the separation increases.

TANK

The right tank choice can have a major impact on the outcome of a plated item. Having a tank that is too small can lead to numerous problems as well as being difficult to use. There are some limiting factors on size of tank, however, the main one is cost efficiency of the tank itself, the electrolyte and electrical costs of heating, agitation and filtration on a large scale.

Material Considerations

When choosing the right tank for electroplating, material selection is critical. Most industrial plating tanks are made from either steel or polypropylene. Having a fully metal tank, however, would result in poor plating and pose electrical risks to the operator. Steel tanks are therefore lined with a non-conductive, chemical-resistant material. Natural hard rubber, neoprene rubber, polyethylene, polypropylene and other plasticized vinyl chloride polymers are materials that are suitable to line plating tanks and be in contact with the electrolyte.

When choosing a tank for plating, there are a few questions that can help with the decision.

4.2: Polypropylene vats are occasionally used in industrial systems for the same reason as the above materials, they are chemical resistant. When designed with reinforcements, polypropylene vats can also hold larger quantities of electrolyte and are much lighter than steel.

What is the tank made from?
Is it chemical resistant?
Can it handle the temperatures?
Is it strong enough to withstand the pressure of electrolyte?
Is it big enough?

4.3: Polypropylene tubs are low cost and readily available, and make for ideal plating, cleaning and rinsing tanks.

4.4: A suitable glass tank for plating larger items. The sealant material was initially tested by placing a small amount of solution in the tank for twenty-four hours.

Coated steel and polypropylene, the two main industrial choices mentioned above, would be suitable to be used in a home environment. That being said, the most readily available, and most cost-effective, option is a polypropylene tub. These can be bought online from many retailers and come in all shapes and sizes. They are chemical resistant and will offer enough strength to hold as much electrolyte as you would need on a small scale. Another good option would be to use a glass tank. Having glass as the main body has multiple benefits: it is chemical resistant; it can withstand higher temperatures (although rapid changes in temperature could cause problems); it is transparent, meaning that with some electrolytes you can watch the process; and it has the strength to hold larger quantities of solution. If choosing this kind of tank, be mindful to check what material the sealant is made from and if this will be attacked by the electrolyte chemicals. Finally, metal tanks can be used if the inside is coated with some of the materials mentioned above. There are numerous sprays available to coat

the metal yourself, so if you are looking for a more robust tank, metal may be the way forward.

Avoid Adhesive

One thing to avoid when lining a tank is choosing an adhesive-backed material. Polypropylene adhesive-backed sheet is readily available and may seem like a good choice, but it will never give a perfect seal. Electrolyte will make its way to the metal by attacking and dissolving the glue. Ultimately, this will ruin your electrolyte!

Size and Shape Considerations

The shape and size of the tank will depend upon what you want to plate. As mentioned earlier, it is better to have a tank that is slightly larger than your requirements. It may cost slightly more but this will ensure that tank size-related failures will not occur. Think about the objects you would like to plate. What is the largest object? What is its overall shape? Is it cylindrical? Is it roughly square? Once you have these in mind then think about what else needs to be in the tank: anodes and anode bags, heater, filter and agitation. All these pieces of equipment take up space in the tank and therefore must also be accounted for when choosing the right size. Now let's consider the amount

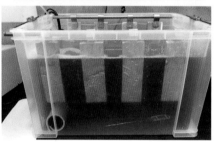

4.5a and b: This large, reinforced PP tank is ideal for large objects or a large number of smaller objects. Not only is there enough room for the parts but there is ample space for heating and filtration equipment and enough anode-cathode separation to give a more uniform current density across the entire surface area. The larger amount of electrolyte will also be less prone to contamination, but the filtration requirements will be larger.

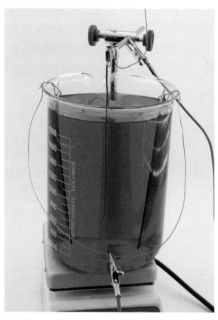

4.6: This glass container can hold up to 2 litres of solution, making it ideal for occasional plating of smaller items.

range is very small and temperatures outside it will be detrimental to the final surface finish.

In a home plating environment, such as a garage, shed or workshop, most plating solutions will happily operate at a small range beginning at room temperature; ideally between 19–30°C. This range works for the most common types of home plating solution: acid copper, acid zinc and Watts nickel.

Having your electrolyte at temperatures lower than the above will mean that the metal salts and crystals will dissolve much slower when first creating your solution. This will require much more mixing until the solution becomes homogenous. When the temperatures are low, plating is slow. A lower temperature means each of the individual particles and molecules will have less energy, so they move slower and result in slower plating and increased electrical resistance. Ultimately, this means you must use more energy to achieve the same thickness when compared with a higher-temperature solution.

The results from a slightly heated electroplating solution, within the home plating range, will be much better. It will initially make the mixing of the electrolyte much easier. Solutions in this range have a higher conductivity, leading to a lower power consumption. Plating will be faster and overall, the final finish will be brighter, more uniform and lustrous.

Operating at higher temperatures can be more detrimental to the results of some plating systems. If the temperature exceeds 40°C, the organic bright-

of electrolyte needed. This can be the most expensive part of the set-up, so it is best to optimize tank shape and size to allow for the minimum amount of electrolyte possible.

HEATING

Heating is an important factor in the results of an electroplated item. In some circumstances, the temperature

4.7a: An example of the results of incorrect temperatures is the brightness of zinc. Using acid zinc at temperatures within the usual range will give shiny, lustrous and levelled deposits.

4.7b: Operating the solution below 15°C will result in dull and patchy finishes.

detrimental temperatures is in Watts nickel solutions. As temperatures rise out of the optimal range, hardness and tensile strength reduce. One downside from a safety aspect is that overly high temperatures can cause the creation of excess fumes and vapours.

Types of Tank Heaters

4.8: Higher temperatures used with copper plating will affect the brighteners used. This will cause the copper deposits to look rough and pink. Higher temperatures can also affect mechanical properties. (4.8b)

Immersion Heating for Large Tanks

Most commercial electroplating immersion heaters are designed for very large plating tanks. For home plating, these heaters can be too large or too overpowered, delivering a very rapid heating effect to a small plating system. They are mostly made of titanium and often have extra components such as liquid level sensors, which are surplus to requirement. Similar types of heaters can, however, be used on a smaller scale.

The most common type of electroplating immersion heaters are stainless steel loops. These are simple heaters that consist of a certain number of loops or coils that are immersed in the plating solution. A few variations of these can be seen in Image 4.9.

These have the advantage of being lightweight and easy to use; simply immerse in the solution to the correct level and turn on. There are usually temperature sensors fitted to these heaters but if not then checks must be made and the heater switched off when the right temperature is achieved.

1.8b

eners and levellers in an acid copper and zinc solutions can begin to denature and can be destroyed with further increases. The effects of high temperatures on an acid copper solution can be seen in Image 4.8. Another example of

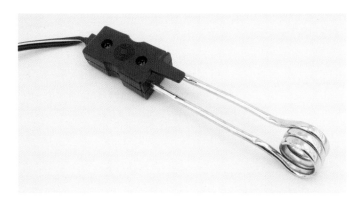

4.9: Steel immersion heaters can be inexpensive and are commonly used. Care should be taken when using them, however, as the metal may corrode in some solutions.

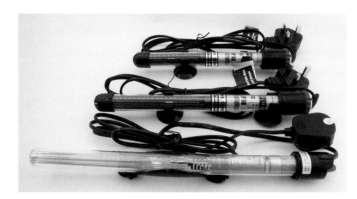

4.10: Glass immersion heaters can be ideal. The three sizes in this image are 100W (top), 200W (middle) and 300W (bottom). The smallest size will happily heat tanks up to 130 litres, with the larger size suitable for tanks up to 350 litres.

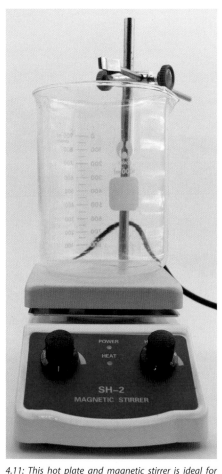

4.11: This hot plate and magnetic stirrer is ideal for mixing up to 2 litres of cleaning or plating solution. Its heating and stirring capabilities make it ideal for the small-scale home plater by reducing the amount of equipment needed.

Tubular Heaters

Tubular heaters are another common form of heater in the electroplating industry. These heaters consist of an outer tube, similar in shape to a test tube, with a heating element inside.

The tubular bodies are made from five different materials: glass, polytetra-fluoroethylene (ptfe), steel, stainless steel and titanium.

Small Tank Heaters

Smaller tanks can suffer from rapid temperature changes due to the small volume of liquid held. This also means that, with the right insulation, minimal power input is needed to achieve the desired temperature. Insulating padding on the outside of the tank coupled with the use of fume control balls can greatly increase the heating efficiency and reduce temperature loss of small (and large) tanks.

If using a very small glass tank, hot-plates can be used. These are efficient and easy to control. An added benefit is that some hotplates have built in magnetic stirrers that reduce the components needed in your plating system.

Heated water baths, often used in lab settings, can also make a good choice of heating system. If the stainless steel inner tank is coated with the right materials, this can also couple as the plating tank.

Glass Heater Safety

Using a glass heater incorrectly can cause a safety issue. The glass can crack and cause a short in the system.

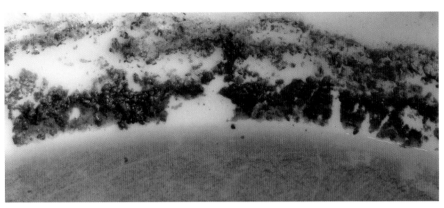

4.13a: *Inorganic materials, usually metallic, can come from a larger number of sources and take the form of larger particles and materials. In this case they come from the part to be plated.*

Filtration System

It is good to keep in mind that, for trouble-free plating, the filtration system used should be able to cope with the heaviest workloads that you intend to undertake. It is also good to remember that, in practice, the number of contaminants that are introduced into the tank are not added at a steady rate. In fact, most contaminants are introduced the moment the work is immersed in the solution. The filtration system you choose should be able to cope with this fluctuation.

4.13b: *They can also come from decaying anodes, as shown, or from precipitation within the electrolyte, and from the surrounding environment.*

FILTRATION

Filtration is another important aspect of getting the best results from your plating tank. A build-up of dirt can, and

4.12: *Organic materials usually take the form of wetting agents, left on parts as part of the cleaning process, or grease and oils such as the machine grease pictured here.*

will, eventually lead to pitting, peeling and dull plating; all things that need to be avoided!

When choosing a filter and filter media, it is important to think about three key points:

◆ Dirt load: soil size, amount, type?

◆ Frequency of filtration: batch or continuous?
◆ Internal or external?
   ◆ Internal filters are more compact and can be cost-effective but can reduce tank space, impose electrical risks and will have a smaller filter size

4.14a: Using a 15× magnifying lens you can see the filter medium from an alkaline tank

4.14b: The filter medium from an acid tank.

The types of dirt, also known as soils, will be introduced in the next chapter but we will just mention the basic types here and where they might come from. There are two classifications of soils: organic and inorganic materials. As a general rule, alkaline solutions will have slimy or flocculent insoluble materials that are difficult to filter. In comparison, acidic solutions contain more discrete, gritty particles that are easier to filter with a course filter media. Oily, slimy organic contaminants will quickly cover the filter media with a dense film, slowing solution flow. In comparison, larger inorganic particles will build a thick layer around and partially within the filter media, but this will only very slightly affect the flow rate. This means that slimy contaminants would require filters with a larger surface area, whereas larger particles would need less. The majority of organic soils are not removed by filtration alone, but by adsorption on activated carbon. This will be covered in later sections.

Sampling Soils

One way to test the type of soils and solids present in your plating tanks is to take a small sample and filter it through white material. The soils should be visible, giving you a good idea of the types present and the most suitable filter media needed.

♦ External filters can be serviced easier, have a larger dirt holding capacity but are often more expensive

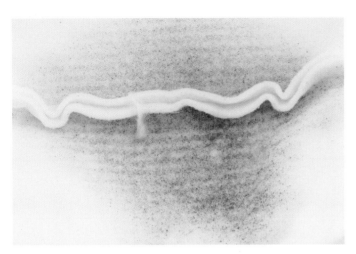

4.15: This filter material has shown that the main contaminants in the solution are inorganic metal particles.

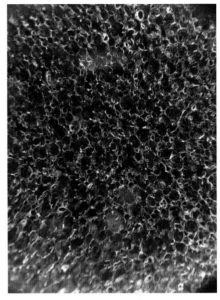

4.17: Coarse filter media, with a 100μm rating, under a 15× magnification microscope.

4.16: From left to right are some easy to obtain filter materials that can be used for home plating: J cloth, 100μm fish tank foam, 50μm car sponge.

4.18: Extra fine filter media, with a 1μm rating, under a 15× magnification microscope.

Filter Media

Fluid Filter media are given a micron rating. This is a generalized way of indicating the media's ability to remove contaminants by directly relating the size of pores and holes to the size of the particles allowed to flow through. The size scale is in micrometres and can range anywhere from a typically coarse media, 100μm, to a dense media 5μm. To give a sense of scale, the diameters of an average human hair and a red blood cell are 70 and 8 microns respectively. The media will also have a percentage rating; this is a measure of the percentage of particles of the given size the media can capture. There are two popular ratings, nominal (50%) and absolute (98.7%). As an example, a filter media that has an absolute 100 micron rating would be able to stop 98.7% of particles that are 100 microns or above. It is important to know that the quantity of particles that can be held by the media is directly proportional to the rating. That is, as the size rating decreases, the number of solid particles that can be held within the media decreases. Ultimately, this means that you will have to change a finer filter more often.

Particle size of debris to be removed is determined by the type of electrolyte. All types of nickel solution will require filter media with a rating between 5 and 15 microns; acid zinc should be between 10 to 20 microns; alkaline zinc from 15 to 75 microns; and copper 1 to 10 microns. Filtration systems always work better when they are oversized. Having multiple filters with a 5 micron mesh will work better and need less maintenance than a single 1 micron filter.

Carbon Filtration

During their lifetime, all plating and cleaning solutions will need to be filtered through carbon. The adsorption of organic molecules and impurities on activated carbon is also critical in waste disposal. Solutions that contain wetting agents, such as acid zinc and semi-bright nickel and most other electrolytes with additives, will need more carbon filtration due to the emulsification of oils and natural build-up of

4.20: Carbon pellets can be added into the solution and left for twenty-four to forty-eight hours to absorb and remove many organic contaminants.

4.21: Agitating the solution will increase the rate at which the carbon will absorb contaminants. Once carbon filtration has been completed, the solution is filtered back into a clean container, removing the carbon and making it ready to use again.

4.21b: Clean solution.

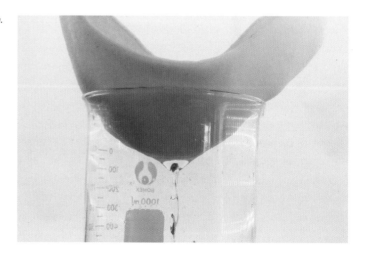

4.19: Carbon for filtration can come in various forms but the most common and useful are carbon pellets.

organic solids. The solutions without wetting agents will need less carbon filtration but skimming instead as oils and grease will form a surface film on the solution.

For home plating applications, carbon filtration will only be needed after prolonged, continuous use or if there are an abundance of contaminants in the tank. The easiest carbon filtration method is through batch filtration. In this process, the solution should be in contact with activated carbon for one hour, with continuous agitation, before it should be finally filtered back into the plating tank. Continuous carbon filtration can become expensive when compared with the more cost-effective batch method.

Anode Bags

Filtration media does not have to be contained within specific pieces of equipment. It can often be very useful to wrap anodes in porous materials, such as PVC and PP cloths. These are known as anode bags. These bags, with the correct ratings, allow for the flow of dissolved ions and electrical current whilst collecting any larger particles, precipitates and contaminants that are created by the anodes. During plating, anode bags will greatly reduce the need for heavy filtration and the amount of typical plating sludge created.

Anode bags are readily available on the internet, but it is also relatively easy to make your own. Bio diesel filter bags, when cut to the correct size, make great anode bags. They are normally made from polypropylene or polyester felt and can be bought in varying micron

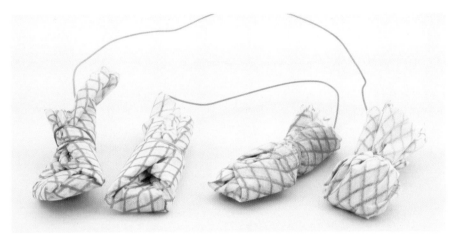

4.22: Nickel anodes wrapped in J cloth and secured with a rubber band to form simple and cost-effective anode bags.

4.23a: The process of heat cutting and sealing anodes bags with soldering iron. Applying pressure onto the metal ruler and soldering iron melts the PP cloth and seals the edges, forming an anode bag. Make sure this is done on a suitable work surface, so the iron does not cause any damage.

4.23b: The edge of the anode bag will be joined together with no holes or gaps that would allow for the flow of soils.

4.23c: The final product, a sealed and ready to use anode bag.

sizes. The plastic material means that the bags can be cut and sealed with heat to create the right shape. Cotton

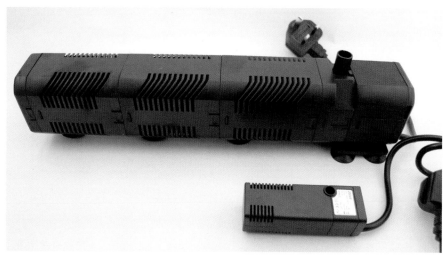

4.24: *Internal filter types. The smaller filter is ideal for smaller plating systems where the levels of filtration and agitation needed are low. The larger filter is more suitable for larger plating tanks or cleaning tanks that need constant filtration.*

Recycling Anode Bags

It is possible to wash out and reuse anode bags. Carefully remove the anodes from the bags and remove the bags from the solution. Allow them to drain. The solution drained from the bags will be loaded with anode particles and will need filtering before being returned to the plating tank. Rinse the anode bag with purified water to remove any remaining electrolyte. Open the anode, turn it upside down and then wash from the outside. This will dislodge and remove particles with the help of gravity. To reduce waste, it is a good idea to filter and reuse any water or solution used in this process.

---

and nylon anode bags should be used with alkaline-based plating solutions. Polypropylene and polyester can be used with both acid and alkaline plating solutions, as can viscose, but viscose can start to break down quicker in highly acidic conditions.

Filtration Frequency

When thinking about tank filtration, it is important to consider how often the tank needs to be filtered. Obviously, if steps are taken in the cleaning and preparation stage to reduce the amount of soil introduced into the plating tank, this will reduce the need for filtration. However, there are some circumstances and types of plating that will introduce unavoidable contaminants. The two main filtration modes are:

- *Continuous filtration* makes use of internal and external pumps that continuously filter the plating solution before, during and after plating. This is the most effective method to maintain uniformity in a plating solution over time. An example of the need for continuous filtration is the plating of bright nickel. In this plating process, organic by-products are generated that will accumulate and cause problems. When sizing your continuous filtration system it should be able to filter twice the volume of the tank per hour.
- *Batch filtration* is needed in the cases where there is either minimal build-up of soils, making continuous filtration needless, or where there is a large amount of soils within the tank that a continuous filtration system could not handle. It is the fastest way of cleaning the solution and involves transferring the contents of one tank into another through a filter media.

Filter Systems

Most filtration systems have three main components: a pump to circulate the solution, a plastic housing containing the filter media and tubing for the transport of the solution. There is an abundance of filtration systems available, not only for the plating industry but for any application that requires fluids to be filtered.

Internal Filters

Internal filters can be compact in size and stick to the side walls of a plating tank. An added feature of most internal filters is that they can act as a source of jet agitation through the circulation of solution in, through and out of the filter. Coupled with the right filter

4.25a: This is an example of an external filter system that can be used with plating.

External Filters

External filters are ideal for medium to larger tanks that have a high plating output. External filters offer high flow with a high level of filtration and much larger capacity than their internal counterparts. This larger capacity can lead to one filter having all the desired filter media, coarse and dense foam and carbon, eliminating the need for batch filtration. External filters also have the added benefit of being easy to access, clean and maintain.

4.25b: The solution is pumped in through the left nozzle, is filtered and is then returned to the tank through the right nozzle, creating a flow of electrolyte around the tank.

Fish Tank Equipment

Home electroplating filters, as well as tanks, heaters and pumps, can be purchased from aquatics and fish-keeping retailers. Many of the variables associated with keeping tropical and marine fish are the same as electroplating and so the equipment is similar. The only consideration to be made when using fish tank equipment is its chemical resistance and the need to make sure that it will be able to be in contact with electroplating solutions for an extended period. If so, then this can be an incredibly cost-effective solution.

media, these can be an ideal solution for smaller tank systems.

Using Anti-foam

An additive called *anti-foam* may be needed when using agitation, especially when using air agitation. Anti-foam controls the amount of surface foam produced by the effects of agitation. In large-scale plating systems, thin foam layers produced on top of the plating tank help and reduce costs by insulating the tank and so reducing heat loss. It also reduces evaporation losses. With home plating, excess foaming can cause a problem. It makes it harder to check the progress of plating and positioning of the item in the tank and it also increases the possibility of touching the bus bar, which can introduce copper contamination and reduce current.

AGITATION

Agitation within the plating system has evolved over the years. When plating first began, agitation was used to prevent stagnation of the electrolyte. In modern electroplating, it has been developed to aid in two features of electroplating: solution mixing and interface agitation. Solution mixing is literally the mixing of the electroplating electrolyte, ensuring homogeneity, the dispersal of gases and the mixing in of any additions to the electrolyte (such as topping up a depleted brightener). Interface agitation is agitation focused on the anode and cathode, specifically in the interface between the electrolyte and electrode (*see Metal-Solution Interphase* section). This specific agitation helps increase the efficiency and effectiveness of electroplating by decreasing the size of this interface and by moving and replacing materials needed for electrochemical reactions. These two features ultimately lead to several benefits in the process:

◆ Avoidance of electrolyte stagnation and dispersal of reactive materials
◆ Increase in deposition rates
◆ Dissipation of heat
◆ Assistance in composite plating
◆ Control of metal deposit properties (grain size and lamination)

Mechanical Agitation

The first type of agitation in electroplating systems was mechanical. This process involved the stirring of the electrolyte by hand as motors had not yet been invented. At the time, this helped early electroplaters to mix electrolyte and avoid stagnation, as mentioned above. When the motor was invented, it was introduced into electroplating in the form of motor-driven stirrers. Rotating the work can be quite impractical and so most stirrers often consist of rotating shafts with propellers at the end. These propellers, when immersed in solution on their own, create a circular, laminar flow of electrolyte around the tank. As work is introduced the electrolyte, flow is interrupted. It acts as a baffle, creating eddies and increasing agitation. There are various standing lab stirrers available that have varying speeds and propeller sizes.

Two major features to watch out for when using electrically driven stirrers are potential corrosion over time and the introduction of contaminants, and the potential for the propeller to strike the cathode. This can cause quite a lot of damage, both to the part being plated and to the stirrer itself. It is therefore best to either use a stirrer in a large tank or to ensure that there can be no contact between the equipment within the tank.

Cathode Movement

For small tanks and for quick, decorative plates, little agitation is needed. This can be achieved by simply wiggling the cathode in the tank during plating.

Air Agitation

Air agitation is relatively common in the electroplating industry, not only for electroplating but for cleaning and rinsing. This is because it is a relatively simple method of mixing and agitation, providing great material transportation that requires little control and initial set-up. In addition to its simplicity, air is free. The only cost is the running of the pump.

When creating air agitation, it is important to choose your equipment

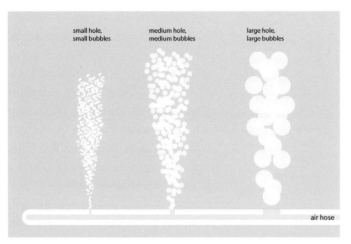

4.26: Varying sizes of bubble. The most ideal air agitation system creates a uniform block of medium-sized bubbles that move laterally as they exit the source.

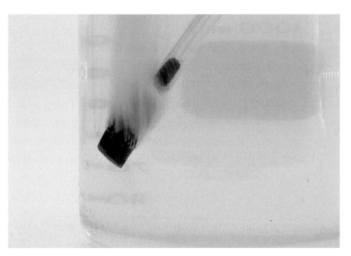

4.27: An air stone such as this will produce enough agitation under a cathode to allow for optimal plating efficiency.

4.28: PP air hose, cut into a ring and drilled with holes, will create similar levels of agitation to that of an air stone. However, the hole size should be small to allow for increased control of agitation.

based on bubble size. As a rule, the bigger the bubble size, the more agitation is created. A balance must be struck, however, as large bubbles will create large disturbances at the surface, adding to the production of fumes and vapours.

There are a few ways of creating the right size of bubbles; this can be with an air stone (make sure it is not corrosive in the solution) or by creating a ring of pipe and manually puncturing small air holes.

There are some disadvantages of air agitation. The first of these, and probably the most significant to the home electroplater, is the production of fumes and vapours. No matter the size of bubbles, as they exit the pipe in the tank, they will grow. This is due to the decrease in pressure and a conglomeration effect where bubbles contact and join with each other as they rise. Larger bubbles will rise through the tank and breach the surface of the electrolyte, popping and causing solution to be propelled into the surrounding environment. This causes vapours, fumes and electrolyte loss. The vaporization also leads to corrosion of ventilation systems and other equipment in the vicinity. It also, more importantly, increases the environmental hazards to the operator and surrounding area. One way of reducing the formation of vapours and fumes are with plastic balls (known as fume control balls, croffles and truffles). These balls can reduce vaporization by up to 90% but do not eliminate it fully. Fume control balls also help to reduce the amount of foaming at the surface. Electrolytes, such as zinc, tend to produce large amounts of

4.29: Fume control balls floating on the surface of an electrolyte. These balls are made from polypropylene, the same material as the tank. They are hollow, and not only do they reduce fumes drastically but also act as insulation to counteract evaporation and heat loss.

4.30: The production of foam can be drastic for some electrolytes. This zinc solution shows the extent to which air agitation can be problematic.

4.31: Antifoam has been added to this solution while an air stone has been used to agitate it. The amount of foam on the surface has been greatly reduced.

Even though the air is only in the tank for a few seconds, this is enough to oxidize materials, promoting the formation of sludge. An example of this is a solution of tin in its stannous form. If agitated with air, the tin will oxidize and fall from the solution.

While these disadvantages have caused many industrial electroplaters to choose alternative methods of agitation, air remains an effective method for smaller, home-scale tank set-ups. Air agitation may require more power, but it can increase the rate of metal deposition.

Vibratory Agitation

A possible alternative to air agitation is through vibratory agitation. Ultrasonic agitation is an example of this. While vibration does not necessarily increase the rate of deposition, it allows other processes to occur that aid in the efficiency of a plating tank. The main benefit is the increased buoyancy given to particles. The vibratory action keeps particles in suspension that otherwise may often separate and fall to the bottom. The vibration also helps to dislodge hydrogen/oxygen bubbles and particles from anodes and cathodes respectively. This indirectly increases plating rates by increasing anode dissolution and creating a larger area for metal to be deposited through the removal of bubbles.

The main advantage is that no air is used. No oxidization occurs and no insulative materials are inserted into the electrolyte. Most importantly, fume and vapour production rates are massively reduced.

foam at the surface when any agitation is used, and especially with air agitation. Excess foam is produced due to the organic materials of the brighteners and levellers. The results can be similar to those seen in Image 4.30. Another additive can be implemented in cases where foam production is high, antifoam. This substance works to reduce the surface tension and drastically reduces foam production.

In the natural environment, air converts metals to metal oxides. The same occurs within the electroplating tank.

Ultrasonic tanks can be used for vibratory action if adequately insulated with non-corrosive materials, as mentioned in the previous section, *Material Considerations*.

POWER SUPPLIES

Electroplating power supplies can be expensive and so the search for an inexpensive alternative can lead to a few problems. Firstly, the most important feature to check for is the safety rating and standards. Buying a power supply that does not pass industry standards can lead to some serious faults in plating and can also cause serious injury to the operator.

Once you begin to choose a power supply you need to make sure that it is suitable for your application. Think about the type of plating you are going to be undertaking: is it manual? Barrel? Rack? PCR?

Manual, barrel and rack plating all require a direct current; the power supply must be able to supply a constant current at a specified amperage. If you are plugging the power supply into a mains socket, it needs to have an inbuilt rectifier, or one must be added to the circuit. It is also important to be able to have a variable output. In almost all home plating circumstances, the size of the objects or the amount in the tank will change, and thus the power output will also need to change. Having a power supply with an inbuilt current controller is extremely useful. This adjustability is usually coupled with a display that tells you the output in volts and amps. Having a power supply without this ability can cause problems but can still

4.32: Make sure to check for stickers like this on any power supply you buy. This will mean that it has been tested after production and that everything should be working correctly.

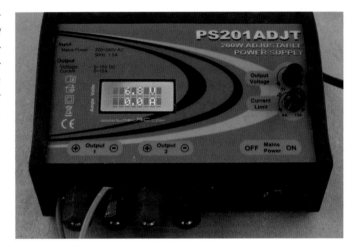

4.33: Having a power supply with a digital readout is extremely handy, especially when plating various items or using different solutions.

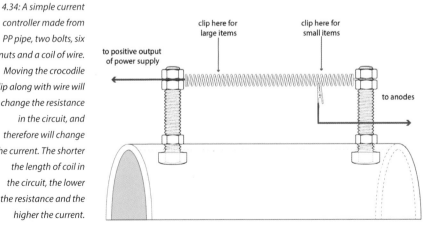

4.34: A simple current controller made from PP pipe, two bolts, six nuts and a coil of wire. Moving the crocodile clip along with wire will change the resistance in the circuit, and therefore will change the current. The shorter the length of coil in the circuit, the lower the resistance and the higher the current.

4.35: *Using a battery charger can be an alternative to a regular power supply. Extra components are needed to achieve specific amperages, such as current controllers. Some battery chargers may not work properly as they have current detection systems that stop the flow of electrons when put into a plating system, making them useless for electroplating.*

4.36: A circuit diagram for a basic electroplating cell.

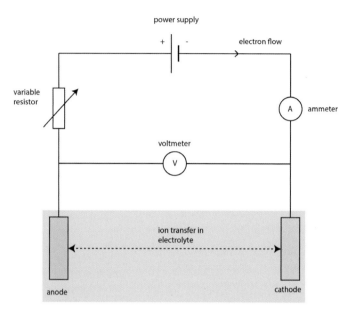

be used if an external current controller is added. If your power supply does not have a current controller then you can use a rheostat, an adjustable resistor, to control the current reaching the work. Alternatively, you can use thin wire wrapped in a coil as an adjustable current controller.

One source of power that can be utilized for home plating is a car battery. When coupled with the previously mentioned ways of controlling current, this can make an incredibly cost-effective means of powering your system. There are obviously more electrical risks associated with this DIY solution.

If periodic current reversal plating, then you will need a power supply that has a programmable rectifier. This will enable the output to be DC, pulse or periodically reversed. Some of these power supplies/rectifiers can be controlled via a computer or will have controls on the front, allowing for the programming of the frequency of the modulated current. The critical parameters of control are the type of regulation, pulse on/off time, cathodic/anodic time and output amplitude.

The next aspect of choosing a power supply is thinking about the power it can supply. You need to make sure it can deliver enough volts and amps to plate your part correctly. The amount of current, amps, you need is explained at the beginning of Chapter 6.

Circuit Set-up

When thinking about power supplies, the obvious extension is to think about the electrical circuit. In addition to having a power supply and anodes you will need wires to connect them all. Make sure they are rated to carry the current that you want to supply. Two useful pieces of equipment to have are a voltmeter and ammeter. Even if your power supply tells you its output, having ways of reading voltage and current at certain points within the circuit can be very helpful, especially when troubleshooting problems.

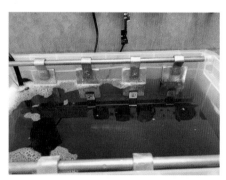

4.37: *The placement of anodes in this set-up allows for a large amount of metal to be deposited. They are evenly spaced and will give good metal coverage.*

4.38: *These copper anodes have begun to dissolve due to prolonged exposure to the electrolyte. As you can see, the lighter areas that have been in contact with anode bags have oxidized little whereas the interior portion, which has been exposed to more electrolyte, has been oxidized and dissolved as the solution has evaporated. Copper sulphate crystals have also begun to grow and spread to other parts of the circuit.*

4.39: *The growth of copper sulphate up from the tank has led to the corrosion of the steel crocodile clip, not only destroying the connector, but also causing a potential safety hazard.*

ANODE PLACEMENT

Anodes should be placed at equal intervals around the tank. This will create a uniform current density on the surface area of the work, which will lead to a more uniform metal deposit. Leaving the anodes in the tank for too long will result in natural dissolution of the anodes by the solution and, in most cases, the growth of unwanted crystals. This growth can spread to other parts of the circuit if connected.

RINSE TANKS

One final piece of equipment needed for electroplating is a rinse tank. In fact, it is good practice to have various ones. These do not have to be as large as regular plating tanks and they also do not have to have the same properties except chemical resistance. The most cost-effective material to use to make a rinse tank is polypropylene. The same shape and size considerations must be made for a rinse tank; they must be able to contain all the parts you have just plated.

5 Cleaning and Preparation

The most important process in electroplating is cleaning and pre-plating preparation. In this chapter we will cover different types of cleaning and pre-treatments, from solvents and alkaline cleaners through to electro-cleaners, rinses, pickles and activators. All these cleaners and processes can be used to achieve a clean and ready to plate surface. Without exception, all metals require some form of pre-treatment prior to plating. The cleanliness of the metal surface can have a dramatic effect on how the metal plates; anything from traces of oils and greases to oxidization to soils and smuts will influence the process and how long the plating will last. Cleaning and preparation is often overlooked and considered of lesser importance than the plating process itself. However, the results of a thorough programmed approach to cleaning and preparation can always be seen in the quality of the final plate.

The most common soils are grease (oil) and oxides (rust). Greasy or oily items will leave a film on the surface of the plating tank, meaning that each time you put something into the tank a layer of oil will cover the part you want to plate. Even small traces can do this, resulting in delamination of the plate. Oxides, rust for example, will cause an item to become over-plated, leaving small nodules on the surface.

5.1a: An example of good cleaning through the complete removal of all surface soils.

5.1b: An example of bad cleaning where grease left on the part has caused skip plating.

Any area of oxidation will grow, even underneath a plated layer, and this will form a blister on your part. Some of the rust will be dissolved into the electroplating solution. High levels of iron can affect current density of a part and therefore deposition rate of the metal to be plated. High amounts of rust can also lead to black spots. If parts are not cleaned properly, the tank can build up high levels of sludge, grit and impurities. These will all negatively affect the way you plate. Tanks that are contaminated will need to be filtered and dummied (a

5.2: Organic soils usually cause problems such as these; dark spots or delamination that can be seen in the top right corner. The black marks came as a result of rinsing in contaminated water where it has left traces of either oil or organic additives.

5.4: This old axe head has a large amount of corrosion that has caused not only oxidization but pitting across the entire surface. It will need to be cleaned using a number of different methods: electrocleaning, mechanical cleaning, and soak cleaning.

5.3: The extent to which the removal of inorganic soils is necessary can be seen in this image. Plating is poorly adhered, leading to blistering, cracking and delamination.

low-current plate that pulls out heavy metal ions in the solution).

TYPE OF SOILS (DIRT)

There are numerous types of cleaning methods possible when it comes to plating and the application of two or more processes is very common and almost always required. The cleaning process that is needed is dependent predominantly on the base metal, but also upon the type of soils on the metal, adhesion to the surface and amount. For example, a pitted, rusty, greasy kickstand will need to be both cleaned abrasively and electrocleaned. Soils can be defined as any unwanted material present on the surface of an object prior to electroplating and are classi-fied as organic or inorganic. Organic soils are usually waxes and oils, whereas inorganic soils can be oxide files and scale, shop dust and dirt and smuts.

The most common soils for industrial electroplaters are:

◆ Mill oil
◆ Forming lubricants – oils used in the metal-forming process
◆ Drawing compounds – lubricants containing molybdenum or graphite
◆ Rust (oxide) preventatives – viscous oils containing organics or soaps
◆ Shop soils – dirt, dust, swarf, cutting oil, inks and fingerprints.

Some of the above can be shared with the most common soils for home electroplaters. The common ones in a home workshop environment are:

5.5: *These metal shavings, swarf from the drilling and sanding of metal, are one example of common shop soils that can contaminate electroplating solutions.*

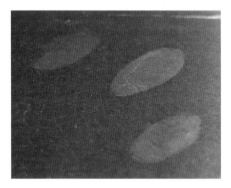

5.6: *If gloves are not worn, touching any clean parts will leave fingerprints such as these. They may not be so obvious, but the grease and skin deposited will cause problems when metal is deposited on top.*

- Shop soils – same as above
- Polishing and buffing residues
- Metallic smuts – metal particles mixed with oils
- Carbon smuts
- Oxides and scale

The best procedure for choosing which cleaning method is to establish first the base material and then determine the types of soil on the surface.

THE BASE METAL

Before beginning the cleaning process, it is important to identify the base metal of the part as this influences the type of cleaning to be used. The condition of the base metal also affects the cleaning process. For example, a piece of new brass sheet would require a totally different cleaning process than a rusty and muddy mild steel bolt. Typically, we can split the cleaning into ferrous and non-ferrous metals, and then more specifically to the types of non-ferrous metals.

STRIPPING

When cleaning and preparing old items, it is important to clean the surface to reveal the base metal. This is to ensure good adhesion as well as reduce electrolyte contamination. On some parts, you will be confronted with the remains of previous plating. This can be dull, faded zinc or multi-layered nickel and chrome with a blistering or peeling surface. Removing this old plating before starting the pre-plating cleaning process is a crucial part in the preparation process. If the stripping is not done correctly the part may be ruined or if the wrong method is employed it can take hours. Also, removing the old plating using the wrong method can be very dangerous.

When starting the stripping process, it is best to remove heavy soils and contamination first. Mechanical cleaning or chemical cleaning, processes that are described later, are suitable for this. This will not only speed up the stripping process but will increase the life of the stripping tank. Remember to

5.7: *Rust and oxides have grown under the zinc surface layer, causing blistering and delamination. Both the oxides and old plating need to be removed before any other process.*

always wear appropriate Personal Protective Equipment (PPE) at each stage of the process. When acid stripping or stripping using electrolysis, make sure it is in a well-ventilated area.

Test it First

When using any stripper or stripping method test first on small areas or non-important items. A small test can avoid any serious damage if the wrong chemicals are being used.

Cadmium

Cadmium is often confused with zinc plating due to its dull finish and colour. Cadmium is harder to strip chemically, therefore people often remove it mechanically. Sanding, shot blasting, buffing and grinding all liberate cadmium from the base metal to create cadmium dust. This dust is toxic and can lead to various respiratory problems. Adequate extraction as well as PPE should be used to stop any such dust from entering the nose, mouth or eyes.

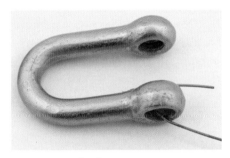

5.8: A cadmium-plated part.

The most effective way to remove cadmium plating is through an immersion process in a 10=20% ammonium nitrate solution. Ammonium nitrate is difficult to obtain, so an alternative solution can be made from 25–50% hydrochloric acid with 16 g per litre of antimony trioxide. Cadmium plating can also be removed electrolytically, the cadmium plated part becomes the anode in a solution containing 200g per litre of ammonium chloride.

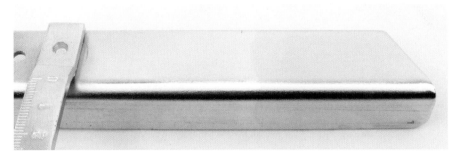

5.9: This part has been plated in chrome. On the right-hand side, the chrome layer has been removed so that the colour change from chrome to nickel can be observed.

Chrome

Chrome is an extremely hard metal and is also difficult to plate; it is, however, easy to remove. The most efficient method is to electrostrip anodically, but there are various solutions and methods that will work well.

The first is to remove the plating anodically using a solution containing 10g per litre of sodium carbonate (soda crystals) and 80g per litre of sodium hydroxide (caustic soda). Cathode current densities of 5–10 A dm^{2} should be used. The second solution should also be used anodically. It contains 50g per litre of sodium carbonate and should be used at a cathode current density of 5–10 A dm^{-2}. The final way of removing old chrome plating is by immersion into a solution. This solution should be operated at higher temperatures, between 20–50°C, and contain 20% hydrochloric acid.

The length of time needed for each of these solutions will vary. When stripping anodically, a yellow cloud will form around the item; this is the chrome being stripped. The main indicator for the complete removal of the chrome layer is a change in colour. In almost all cases, the metal beneath the chrome is nickel. The colour change from chrome to nickel is subtle, but nickel will look slightly more yellow than chrome.

Copper

There are two solutions that will strip copper layers from the most common metals. They are both used anodically and operated at room temperature, around 20°C. The first is a solution of 120g per litre of sodium sulfide. The second is a solution of 190g per litre of sodium nitrate. They should both be used at a cathode current density of 2 A dm^{-2}. Brass layers can also be removed with both solutions.

Gold

Gold is removed anodically with a simple and relatively safe acidic solution consisting of 100g per litre of citric acid – a common food additive. Its operating temperature is, like most stripping solutions, 20°C and it should be operated at 4.5 volts.

5.10: The results of removing old nickel plating can be seen here. All the previous surface has been removed from the part, but it still retains areas of oxidization that will require further cleaning.

hours to fully remove the nickel. It is relatively safe for the base metal as it will not attack the aluminium. Instead, it acts on the nickel and its bond to the zincate strike layer.

Silver

Similar to gold, silver can be stripped anodically. The solution used contains 200g per litre of potassium ferrocyanide and 10g per litre of potassium carbonate. It should be used at 40°C and at a current density of 4 A dm^{-2}.

Nickel

Nickel plating can be difficult to strip as it is applied to numerous base metals. In most circumstances, it is easier to purchase preformulated nickel and nickel alloy strippers from specialist companies. There are, however, some homemade solutions that will work if you know the base metal.

A relatively safe option for the anodic stripping of nickel from most base metals is with a solution containing 500g per litre of sodium nitrate. This will begin to etch the surface of metals such as zinc and iron. This sodium nitrate solution should be operated at high temperatures, between 80–90°C and at cathode current densities of 10 A dm^{-2}.

Removing nickel plating from steel and copper can be achieved anodically, using a solution of 30–50% sulphuric acid and 30g per litre of glycerine. The solution should be at a temperature of 20°C and used at a current density of 15 A dm^{-2}.

Nickel is often plated onto alumin-

ium. Removal from this metal can be completed using immersion processes in proprietary solutions. These often come under such names as 'aluminium deox and desmut solution'. They are normally a mix of between 3–5% nitric acid and 5–10% ammonium bifluoride. This dip method is undertaken at room temperatures and can take several

Zinc and Zinc-Nickel Alloy

Zinc and its alloys are easy to remove. Most acid solutions will strip zinc either through an immersion process or by anodic electrolysis.

The immersion process is much easier as it involves no electricity and only one chemical. A 10–15% hydro-

5.11: The process of immersion nickel stripping causes yellow precipitate in the solution. This is nickel and nickel hydroxide being removed from the part.

5.12: When the item is immersed in the stripping solution you will notice bubbles forming on the old plating. This will increase to a rapid fizz. The warmer or more concentrated solution, the more vigorous the fizzing will be. This process will only take a few minutes and should be monitored closely. When the fizzing slows or virtually stops this is an indication that the plating has been removed.

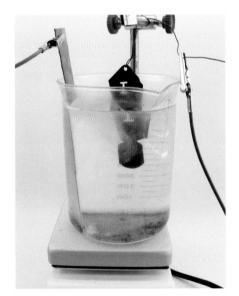

5.13: Electrostripping of a zinc-plated part in an alkaline solution.

		Composition		Temp (°C)	Cathode Current Density (A dm^{-2})
		Material	Quantity (g L^{-1}) (% by weight)		
Cadmium	Immersion 1	Ammonium Nitrate	10–20%	20	
	Immersion 2	Hydrochloric Acid	25–50%	20	
		Antimony Trioxide	16		
	Electrolytic	Ammonium Chloride	200	20	1–2
Chrome	Electrolytic 1	Sodium Carbonate	10	20	5–10
		Sodium Hydroxide	80		
	Electrolytic 2	Sodium Carbonate	50	20	5–10
	Immersion	Hydrochloric Acid	20	30–50	
Copper	Electrolytic 1	Sodium Sulfide	120	20	2
	Electrolytic 2	Sodium Nitrate	190	20	2
Gold	Electrolytic	Citric Acid	100	20	4.5
Nickel	Electrolytic 1	Sodium Nitrate	500	80–90	10
	Electrolytic 2	Sulphuric Acid	30–50	20	15
		Glycerine	30		
	Immersion	Nitric Acid	3–5%	20	
		Ammonium Bifluoride	5–10%		
Silver	Electrolytic	Potassium Ferrocyanide	200	40	4
Zinc	Hydrochloric	Hydrochloric Acid	10–15%	20	
	Inhibited Hydrochloric	Hydrochloric Acid	10%	20	
		Antimony Trioxide	5		
		Stannous Chloride	5		
	Electrostripping	Sodium Hydroxide	10%	20	0.5–3.5

Table 5.1: Stripping solutions.

chloric acid solution is commonly used as this chemical is widely available and cost-effective. If available, inhibited hydrochloric acid can be used. This will make the process more controllable by a reduction in the attack on the base metal. It is possible to make your own inhibited hydrochloric acid solution. Simply dilute hydrochloric acid to a 10% mix, then add 5g per litre of antimony trioxide and stannous chloride. Sulphuric acid or a multi-acid salt solution can also be used, again, at concentrations of 10 –15%.

One problem with the immersion process when cleaning old zinc plating is that if you use a more concentrated acid solution or a hot solution, fizzing will still occur. However, this is from the base metal after the zinc has been stripped. It is important to remove the items once the old plating has been removed because the acid will start to etch the base metal.

Electrostripping of old zinc plating is

done by plating it onto another metal. Using a 10% sodium hydroxide solution, the part to be stripped is connected to the positive end of the power supply and a steel object is made the cathode. Similar current density values to normal alkaline zinc plating should be used.

CLEANING METHODS

Metal cleaning can usually be accomplished with a combination of five different methods.

A) Soak cleaning
B) Abrasive cleaning
C) Solvent cleaning
D) Detergent spray or wash
E) Electrocleaning

One May be Enough

Depending on the initial condition of your part, you may only have to apply one of the cleaning steps to get the metal ready and activated for plating.

What do we mean by a clean surface? Well, some people's definitions may vary, but really, it is the removal of all dirt, oxides and unwanted contaminants that will affect adhesion and finish of the metal surface. In plating terms, a clean item will have an activated surface made entirely of metal particles and no

5.14: Soak cleaning of an axe head in an acidic solution for the removal of oxides.

5.15: Abrasive cleaning using a wire brush drill attachment. This can be the fastest way to remove large amounts of surface soils while slightly smoothing the pitted surface at the same time.

5.16: A solvent cleaner being applied to a soiled part with a trigger spray bottle.

other contaminants, allowing for free flow of electrons that will give a surface with a high current density, permitting reduction reactions to occur. This can be achieved easily, all it takes is a little time, effort and attention to detail.

5.17: Detergent cleaning is ideal for removing some organic and inorganic soils. This part is being cleaning in a warm solution containing 1% sodium lauryl ether sulphate.

5.18: Electrocleaning tank set-up. The tank contains a light-duty alkaline cleaner and the anodes are steel. They are wired either side of the tank to allow for the full cleaning of an item placed in the centre. This process can cause a lot of fumes and so extraction is important.

CLEANING METHOD A – SOAK CLEANING

Soak cleaning is the main type of cleaning process used for plating. It is used to remove heavy oils and greases, quickly, effectively and economically. There are numerous different types of soak cleaners, but they can be classified as either acidic or basic, and heavy or light duty. When using a soak cleaner, the items to be cleaned are submerged in the cleaning solution. The temperature and concentration of the cleaners should be as high as possible for the type of metal to be cleaned without any harmful effects to the metal. This means that the time in the cleaner will be reduced.

Table 5.2: Light-duty alkaline cleaner composition and operating conditions.

Light-Duty Alkaline Cleaner Composition and Operating Conditions		
Chemicals & Operating Conditions	Chemical Formula	Measurements
Sodium Hydroxide	NaOH	5 gL^{-1}
Sodium Carbonate	Na$_2$CO$_3$	22.5 gL^{-1}
Sodium Gluconate	C$_6$H$_{11}$NaO$_7$	2 gL^{-1}
Sodium Metasilicate	NaSiO$_3$	15 gL^{-1}
Sodium Tripolyphosphate	Na$_5$P$_3$O$_{10}$	22.5 gL^{-1}
Borax	Na$_2$B$_4$O$_7$	1 gL^{-1}
Sodium Lauryl Ether Sulphate	CH$_3$(CH$_2$)$_{11}$SO$_4$Na	2.5 gL^{-1}
Temperature		80°C
Time		2–4 minutes

Light-Duty Soak Cleaners

Light-duty cleaners are most often used for non-ferrous metals. The contents are: wetting agents, buffering salts, sequestering agents, dispersants, inhibitors and couplers. Due to the inhibitors in the solution, light-duty cleaners are sometimes referred to as inhibited alkaline cleaners. They work by wetting, emulsifying, dispersing, and dissolving the soils, and have an operating temperature from 60 °C to just under boiling point. Chemical concentrations of 5–10% per litre are normal and the pH ranges from 11 to 12.5, making it an alkaline soak cleaner. Light-duty cleaners are ideal for most non-ferrous metals, like copper and brass, and are especially good for cleaning aluminium and zinc alloys.

Heavy-Duty Soak Cleaners

Heavy-duty cleaners are most often used for ferrous metals. Similar to light-duty cleaners, they include wetting agents, buffering salts, sequestering agents, dispersants and inhibitors. Their composition is similar again to light-duty cleaners, that is, they are

5.19: Light-duty soak cleaner currently cleaning an item. This cleaning solution is at a temperature of 60°C and is formulated from sodium hydroxide, sodium carbonate, sodium metasilicate and sodium lauryl ether sulphate. The tank contains a brass curtain hook that needs grease removing from the surface.

based on highly alkaline substances such as sodium hydroxide, potassium hydroxide, sodium carbonate, trisodium phosphate and tripolyphosphates. The operating temperature has a tighter range, from 70 to 80°C. Chemical concentrations are higher, up 20% per litre for some materials, meaning that the pH is also higher, from 12.5 to 13.5. Heavy-duty alkaline cleaners are good for removing fats, oils and grease as well as rust preventatives.

A problem with this type of cleaner is that occasionally, once the oils and fats have been dissolved in the solution, they can form metallic sodium

Heavy-Duty Alkaline Cleaner

If a higher rate of sodium gluconate is used in the heavy-duty alkaline cleaner it can also be used to remove rust and lime scale. The table below shows the composition of a typical heavy-duty alkaline cleaner.

Chemical Composition & Operating Conditions	Chemical Formula	Measurements
Sodium Hydroxide	NaOH	70 gL⁻¹
Sodium Carbonate	Na_2CO_3	15 gL⁻¹
Sodium Gluconate	$C_6H_{11}NaO_7$	2 gL⁻¹
Sodium Metasilicate	$NaSiO_3$	15 gL⁻¹
Sodium Tripolyphosphate	$Na_5P_3O_{10}$	10 gL⁻¹
Sodium Lauryl Sulphate	$CH_3(CH_2)_{11}SO_4Na$	2.5 gL⁻¹
Temperature		80°C
Time		2–4 minutes

Table 5.3: Heavy-duty alkaline cleaner composition and operating conditions.

soaps. These can be very difficult to remove. Sodium metasilicate is added to combat this problem. The silicates are used to stop corrosion of the base metal and stop these metallic soaps from forming.

Acid Cleaners

Acid cleaners are specifically formulated to remove organic soils and dirt, and thin oxides from ferrous metals. Unlike pickling and derusting acid

5.20a: Before: copper with a fingerprint. The natural grease from the hand has caused oxidation in the form of a fingerprint.

5.20b: After: immersion in an acidic cleaner has removed the oxides and grease, leaving a clean surface ready for plating.

tanks, which contain a high concentration of strong acids, acid cleaning tanks have a higher pH and are designed to remove very light grease, oxide films and organics.

Acid Activators, Pickles and Etches

These are acid treatments that are used immediately before plating. Like acid cleaners, they remove light grease and oils but, most importantly, they activate the surface of the metal by removing oxides and etching the part slightly – leaving the base metal at the surface. This etching will help with adhesion in the subsequent plating process. Acid activators can also be very helpful as they can get rid of difficult to remove oxides, such as those from nickel and stainless steel. In addition to surface etching, activators are also used to neutralize and remove any remaining alkaline cleaner. Acid pickles are used predominantly as a dip, but they can be used cathodically. They also help to neutralize any alkaline left from the alkaline cleaning stage before it can contaminate the plating solution.

Mineral acids are the most commonly used pickle or activator. Often, wetters are added into the solution to reduce surface tension and improve the overall cleaning effect. Mixtures of acid salts are also used to improve the activation effect on the metal surface. Temperature, concentration and agitation will also play a part in their effectiveness – increases in temperature and concentration, coupled with agitation, will decrease the time needed in the activation tank.

There are two key features that you

Ingredients	Steel g/L	Stainless steel g/L	Nickel g/L	Zinc g/L	Copper g/L	Brass g/L	Aluminium g/L
Sodium Bisulphate	20	40	40	10	12	15	40
Sodium Fluoride	5	10	10	2	4	4	10
Sodium Hydrogensulphate	75	150	150	38	45	50	150
Operating conditions							
Time in minutes	1-2	0.5–2	1–5	0.1–1	0.5–1	0.2–1	0.1–0.5
Temperature °C	20–30	30–50	40–50	15–25	20–30	20–30	30–60

Ingredients	Steel g/l	Stainless steel g/l	Nickel g/l	Zinc g/l	Copper g/l	Brass g/l
Hydrochloric Acid	10	30	25	5		
Operating conditions						
Time in minutes	1–2	5	1	0.1–0.4		
Temperature °C	20–30	20–30	20–30	15–25		

Ingredients	Steel g/l	Stainless steel g/l	Nickel g/l	Zinc g/l	Copper g/l	Brass g/l
Sulphuric Acid	10		15	0.5	15	10
Operating conditions						
Time in minutes	1–2		1	0.1–0.4	0.5–1	0.5–1
Temperature °C	20–30		40–50	15–20	20–30	20–30

Tables 5.4, 5.5 and 5.6: Composition and operation of acid activators.

need to look for when selecting the type of activator or pickle to use. Firstly, it must not etch the metal surface too much. Second, it must not form a non-soluble metal salt that can affect the subsequent plate.

The tables above show the composition of acid solutions used for cleaning the various metals along the top of the table. Their quantities are shown in grams per litre. There are a few metals that we have left out:

- Lead, leaded steel and leaded brass can be activated by using sulphamic acid 30g/l or citric acid 60g/l, and either can be used at room tem-perature and immersion time of 1-2 minutes.
- Silver is activated with 100g/litre citric acid at 60°C for 1 minute or with a mixture of 20g per litre nitric acid and 90g per litre thiourea at room temperature for 1 minute.

Ultrasonic Cleaning

Ultrasonic cleaning is another soak cleaning process where the metal is submerged and left inside a tank that then uses sound to clean the part. The action that cleans the part is very similar to electrocleaning, which will be described later in this chapter. Essen-tially, the sound waves produce micro-bubbles on the surface of the metal that collapse and re-form, ripping away soils to do most of the cleaning work. This bubble formation and collapse is called cavitation. The pressures and tempera-tures created are extremely high, which is why it is such an effective cleaning method. One great benefit of ultra-sonic cleaning is that the bubbles form all over the surface, meaning that the smallest of cracks or holes are cleaned or parts with complex geometries are cleaned with little effort. You can use alkaline, acidic, solvent, detergent or neutral cleaners within an ultrasonic cleaning tank. Ultrasonic cleaning can be one of the best cleaning methods available. The only limitation, however, is the expense of the large tank.

CLEANING METHOD B – ABRASIVE CLEANING

This section covers a multitude of dif-ferent abrasive cleaning techniques from media blasting to fine polishing. It can also be used on virtually all types of metal. The benefits of mechanical cleaning can lead to a very short time in the plating tank. As you work to clean the parts, you can very precisely control the amount of material removed from the surface and the overall finish of the parts to be plated. If you were to achieve a pristine surface, then a 5–10 minute plate would be all that is required for a decorative, lustrous mirror finish. As mechanical cleaning is a very manual process, it is usually very labour-inten-sive, meaning that some parts can take a very long time to get to the right level of cleanliness before plating. If you do not want to spend much time physically scrubbing your part, then there are pieces of equipment that can help. These, however, can be expensive to set up and run.

Reduce Dust

With abrasive cleaning methods, a large amount of metal dust can be created. Make sure that you perform any cleaning actions with adequate ventilation or while wearing a dust mask and goggles.

5.21b: Fine material used for blasting.

5.22a, b and c: These images show a rusted and pitted induction manifold with no cleaning prior to a blast cleaning treatment. Oxide, old plating and oil cover the majority of the surface.

5.22b.

5.22c.

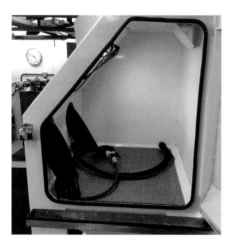

5.21a: The internal compartment of a blasting cabinet. The material is pumped through the system, either with air or liquid, and blasted onto the items. The materials can then be collected and recycled through the grate at the bottom.

Blasting

Sand, media or vapour blasting can be one of the best ways of removing oxides, scales, paints and lacquers. It can be a relatively quick process compared with some of the other ways of removing these types of soils. The downside to sand blasting is that it requires a sand blasting gun or cabinet, a high-volume compressor and the appropriate media. This can be anything from aluminium oxide to fine glass beads, to shell or sodas. Due to the hardware needed and some of the media used, it is quite expensive to set up and run. Another downside of media blasting is that it is powered by compressed air. Even the best compressors, with liquid traps, can still lead to small levels of contaminants

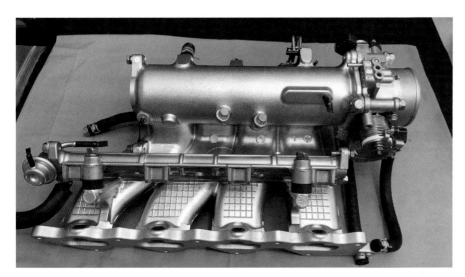

5.22d, e and f: After blasting treatment, the manifold looks immaculately clean. The oxide has been removed and the fastenings have been zinc-plated and passivated with a yellow chromate conversion coating.

(oil and moisture) getting through the system. These are then transferred, via the media, to the item. Sand or media blasting can also lead to impregnation of particles in the surface of the item.

Vapour blasting, which uses a water-based abrasive solution that is accelerated onto the surface of the item by means of compressed air, has a couple of advantages. Firstly, there is no dust produced. Secondly, the cleaning is gentler and more thorough. This is because the water cushions the impact of the abrasive and removes any remaining residues. In theory, good mechanical cleaning can be used as a

5.22e.

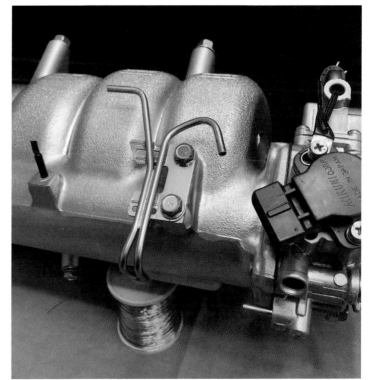

5.22f.

single pre-plating cleaning and deoxidization step. In some cases, this can be followed directly by plating. If you are thinking of using this method of cleaning then always check with a water break test (described in later sections) that no traces of oil or grease are left on the metal surface.

5.23b: Similar removal and smoothing of items can be achieved using wire brushes and wire wheels such as these various drill attachments. They make for an easy way to remove light surface soils and oxides.

5.23a: Sanding media can take a few different forms, wet and dry paper, sandpaper and various Dremel and drill attachments.

Sanding

Sanding using production paper, wire wool or emery cloth can be quite a good choice of abrasive cleaning. It is relatively cheap but it is very labour-intensive. Sanding using a linisher or flapped wheel will take quite a lot of the hard work away but it still means that you have to physically hold and clean the parts. One of the advantages of sanding is that by using finer and finer grades of paper you can go from removing heavy oxidization and paint down to a very smooth, almost polished, finish. If the sanding is done correctly, without the transfer of soils onto the cleaned part, it can also be a single-step process prior to plating.

5.24: Fine wire wool is excellent at removing thin oxides before plating and if the part is clean, it can be used as the activator before plating, providing it passes the water break test.

Polishing

Buffing or polishing is a great way of preparing the surface of relatively clean metal, especially when a mirror finish is required on the plated item. With the use of different buffing mops and compounds, you can clean and polish most types of metal. Buffing can be done using several different types of machines:

◆ Benchtop buffing wheels
◆ Benchtop grinders with adaptors to take buffing mops
◆ Hand drills with the correct tooling
◆ Dremels

When it comes to polishing and buffing heavily pitted, oxidized or painted parts, you would need to start with hard mops with highly abrasive compounds.

The main function in this step is to remove any unwanted soils and level

5.25: Even though this part has been given a thick copper plate, there are large pits that need removing. This can be done through mechanical cleaning: sanding buffing and polishing.

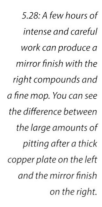

5.28: A few hours of intense and careful work can produce a mirror finish with the right compounds and a fine mop. You can see the difference between the large amounts of pitting after a thick copper plate on the left and the mirror finish on the right.

5.26: There are a wide variety of buffing mops and compounds available. It is best to have three mops: hard, medium and soft and a polishing mop. Having different compounds will also help to better the surface finish.

parts ready for plating it can take anywhere from half an hour to half a day to achieve the right finish. There is also a chance that after polishing residues will be left on the parts and these can be difficult to remove. You could simply place these parts in an alkaline soak cleaner. The benefit is that you can

achieve a mirror-finished metal surface. This means that you only need a thin surface plate to maintain the finish.

5.27: After sanding and buffing with a heavy mop, the difference in the axe head can be seen. Further buffing and polishing is needed to achieve a mirror-bright final plate.

an uneven surface. Once you have removed any major flaws, intermediate mops and less abrasive compounds are used to smooth the surface further before soft mops and light compounds are used for the final mirror finish.

There is one major drawback when it comes to buffing and polishing – it is labour-intensive. To get perfect

Removing all the Residue

When polishing an item, make sure to remove all the polishing residue with extra cleaning steps. Failure to do so will result in blistering and scorch marks on the final plate.

5.29: Heavy polishing compound can be seen on this part; it is the black material surrounding the lighter, recently buffed section.

CLEANING METHOD C – SOLVENT CLEANING

Solvent cleaning is particularly effective at removing oils, grease, waxes, tars, resins and some glues. In industrial cleaning systems, non-flammable chlorinated hydrocarbons, such as trichloroethylene, are the solvent of choice. These solvents, however, are toxic and are not recommended for home use. Safer substances for home platers to use are solvents with low flash points.

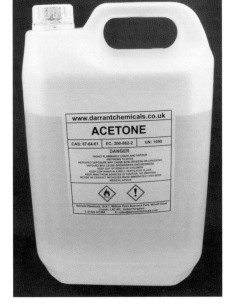

5.30: Acetone, used as a solvent cleaner to remove light oxides, grease and oils. It is best used in conjunction with fine wire wool, brush or another, less harsh, mechanical cleaner/action.

Isopropyl alcohol (IPA) is a solvent that is ideal for home use as it is relatively safe, cheap and readily available. It is good at removing polar or ionic residues – such as soldering flux – and is also good on some oils and grease

5.31a: A greasy nickel-plated part before solvent cleaning.

5.31b: The part has been partially immersed in solvent (IPA) and is being washed with a paint brush.

5.31c: The part has been completely cleaned, with the grease transferring from the part into the solvent.

5.31d: The final cleaned part.

and general handling residues. It is also effective on ionic salts such as sodium, calcium, sulphates, ammonia, chlorine and fluorine. The downside to IPA is that it is not good on non-ionic or non-polar soils. Essentially, organic soils and the organic molecules that make up oils, grease and other hydrocarbons are not dissolved, so when using IPA, use it in conjunction with non-polar hydrocarbon solvent such as pentane.

Another solvent that is often used (especially by 3D printing enthusiasts working with ABS) is acetone. Although it is highly flammable, it is not considered an air pollutant or a volatile organic compound (VOC). It is a great cleaning substance in many ways, but it is an especially effective degreaser and soil remover. It also has low toxicity, is inexpensive, easy to obtain and is water soluble. It is particularly good at removing some types of glues, lacquers and paints when used as a soak.

Wax and grease remover, which is usually a blend of several chemicals often containing xylene, hexane, toluene, acetone, etc, is probably one of the best ways to obtain solvent cleaners for removing oils, waxes and grease. It is also pretty good at removing buffing residue. The problem with this type of cleaner is that it is a VOC, very flammable, toxic and an environmental pollutant.

CLEANING METHOD D – DETERGENT SPRAYS AND WASHES

There are lots of detergent cleaners available on the market, most of which can be used as part of the metal finishing process. Most detergents contain alkylbenzene sulfonates, which are similar to soaps but are more soluble in water. Detergents work because they are amphiphilic; partly hydrophilic (water-loving) and partly hydrophobic (water-phobic). Their dual nature facilitates the mixture of hydrophobic compounds (such as oil and grease) with water. When you only have light oils and/or grease to remove, then detergent cleaning is all that is needed. It is a cheap and easy solution to clean parts and does not need the time or money that other cleaning solutions require.

Detergents work best with some sort of physical action as part of the process. Washing items by mechanically scrubbing in a 1% mixture of detergent and hot water gives effective cleaning of light oils and greases. Another option is to use a sprayer to apply the detergent mixture directly onto the metal surface. This has the benefit of having clean detergent and water contacting the surface all the time while the used detergent and water runs off.

5.32a: A similar part to that of the previous image of solvent cleaning, a dirty nickel-plated ring in need of cleaning and a solution of warm, soapy water.

5.32b: The part is immersed in the warm solution and scrubbed clean with this brush.

5.32c: After cleaning, all traces of soils are removed. The black marks visible are spots of oxidization that has corroded the surface layer.

5.33: With anodic cleaning, oxygen is liberated at the anode in the process of oxidation, the loss of an electron. Specifically, hydroxide ions, OH^-, migrate toward the positive anode, where the react with each other to form water, oxygen and electrons. The reaction is $4(OH)^- \rightarrow 2H_2O + O_2 + 4e^-$.

5.34: Cathodic cleaning creates hydrogen in a reduction reaction where hydrogen ions gain electrons and join to form hydrogen gas. The equation for this reaction is $2H^+ + 2e^- = H_2$.

CLEANING METHOD E – ELECTROCLEANERS

Electrocleaners are often used as one of the final cleaning stages prior to plating. The same heavy-duty alkaline cleaners are the basis of most electrocleaners. They are used with DC currents either anodic, cathodic, or a combination of both known as periodic current reversal (PCR). On relatively clean metal, electrocleaning alone will often be enough to clean and activate the items ready for plating.

Anodic cleaning is the process where the part needing to be cleaned is made the anode (positive). This process dissolves the base metal slightly and therefore should be used first whenever possible. It removes any adherent films and metal particles such as metal smuts and oxides. Anodic cleaning is also brilliant at removing organic soils.

Cathodic cleaning is the reverse of anodic cleaning; the part to be cleaned is made the cathode (negative). At the cathode, reduction occurs. Unlike electroplating, in electrocleaning there is nothing deposited on the cathode; instead hydrogen gas is formed by the reduction of hydrogen ions from the cleaning solution.

5.35: The action of the formation and motion of the hydrogen gas causes unwanted particles to be removed.

This hydrogen gas forms microbubbles on the surface of the metal and it is these bubbles that help with the removal of soils, forcing them free of the surface as the bubbles are formed and rise. This action, in addition to the alkaline solution (as per soak cleaning section), is how soil removal is accomplished.

Hydrogen Embrittlement

An effect of cathodic cleaning is hydrogen embrittlement. This is when hydrogen is not liberated from the surface as bubbles but remains on the metal. Spring steel is one example of a metal that is critical to embrittlement. It can be removed easily by heat treatment immediately after cleaning. Essentially, this process requires one hour in the oven at 200°C.

Alkaline Electrocleaning

The below table shows the operating conditions for an alkaline-based electrocleaner. It describes the current density needed, polarity it can be used at, alkalinity, temperature and time per base metal.

Electrocleaner operating conditions for different metals. Note: a. For barrel operations, a small fraction of current density is used; b. Polarity: A = anodic current, C = cathodic current, PRC = Periodic Reverse current; A-inh = Anodic current with inhibited electrocleaner; c. Alkalinity, as g/l NaOH

Cathode Cleaning

Cathodic cleaning should be used to clean tin, lead, brass and aluminium as well as activate nickel and high-nickel alloys.

After alkaline electrocleaning, items must be well rinsed and then dipped in a dilute mineral acid. This process neutralizes any alkaline films and removes any light or remaining oxidization. Hard

Metal	Current Density[a] (A/dm²)	Polarity[b]	Alkalinity[c]	Temperature (°C)	Time (min)
Steel, Low Carbon	3–5	A, C, PRC	50–100	60–90	1–5
Steel, High Strength	5–10	A	50–100	60–90	1–5
Stainless Steel	5–8	C	50–100	60–90	1–5
Copper	5–8	A, C	50–100	60–90	1–5
Brass	2–4	A, C, A-inh	15–20	50–70	1–3
Zinc Die-cast	2–4	A, C, A-inh	50–70	50–70	1–3
Nickel and Ni alloys	2–3	C	30–60	50–80	1–5
Lead and Pb alloys	5–8	CC	---	50–60	1–3
Silver and Ag alloys	2–3	A, C, PRC	---	50–60	1–3

Table 5.7: Operating conditions for an alkaline-based electrocleaner.

Ingredients	Steel g/L	Zinc g/L	Copper g/L	Brass g/L
Sodium Hydroxide NaOH	50	7	12	7
Sodium Carbonate Na_2CO_3	3	9	12	10
Sodium Tripolyphosphate	4	3	5	4
Sodium Metasilicate	30	13	20	13
Wetting agent(s)	1	1	1	1
Operating conditions				
Time in minutes	2–10	0.5–1	0.5	0.5
Current density (A/dm2)	5–10	2–4	5–8	1.5–4
Temperature °C	80–95	70–80	70–80	60–70

Table 5.8.

5.36: This old vice may look almost impossible to clean, it is covered in rust, paint and layers of dried grease.

5.37: On closer inspection, you can see the extent of the oxidization of corrosion of the parts.

5.38: The thread of the clamp is completely blocked with dried grease and rust.

to remove scales or oxidization may require two cycles of alkaline electro-cleaning. The first longer, electro-clean may start off with a cathodic clean relying on lots of gassing and scrubbing action to lift the scale and rust. At the end of this first cycle anodic cleaning is used to remove smutting. Lower current is used during the first part, while higher currents are used during the anodic cleaning stage. Once the oxides and scale have been removed, the second electrocleaning cycle is just to remove any remaining smutting.

A balanced solution is very important when electrocleaning. To get the best results it is better to alter or adjust the balance of the cleaner depending on the type of metal to be cleaned. The table below shows some common metals and the concentration of chemicals needed for each one. How the metal reacts in the electrocleaner will determine if anodic cleaning or cathodic cleaning is needed. Nickel, brass and stainless steel are nearly always cleaned cathodically. Zinc and zinc alloys are normally cleaned anodically to stop the metallic soaps being deposited. They are also easily attacked by highly alkaline cleaners, so inhibited cleaners with lower levels of NaOH are used. Steel and stainless steel can be cleaned anodically or cathodically. Stainless steel is more affected by anodic cleaning and may require a longer post-cleaning acid dip (pickle).

Anodic cleaning is always preferable where possible. This is because this method will slightly dissolve the base metal, which releases anything on the surface. It also prevents any surface films or metallic soaps sticking to the metal surface. Anodic cleaning can produce thin oxide layers; these oxides are easily removed with the following acid dip. With anodic cleaning, oxygen is produced at the metal surface, which means there is no hydrogen embrittlement. An example of the effectiveness of electrocleaning can be seen in Images 5.36–5.40b.

Cathodic cleaning – sometimes referred to as direct electrocleaning – uses the same cleaners, temperatures and current densities as anodic cleaning but the polarity is reversed so the item to be cleaned is connected to the negative side of the circuit. With cathodic cleaning, twice as much gas is produced at the cathode, so the scrubbing action is much better. It is mostly used to remove heavy soils and deposits, and even rust and paint can be removed in this way.

The downside to cathodic cleaning

5.39: The surface for the alkaline cleaning tank shows just how much soil can be removed by electrocleaning. Not only has all the rust been removed but the layers of paint have been stripped and can be seen floating.

is that any positively charged material in the cleaner will be attracted to the item. This material is normally not very adherent, but it can still be difficult to remove. Also, high-tensile steels such as spring steel or high-tensile bolts with a hardness rating of 320HV, so bolts marked on the head with 8.8 or above or with three radial lines spreading out from the centre of the bolt will require heat treatment after the cleaning and plating process.

5.40a and b: Once the parts have been removed from the cleaning tank and rinsed the brilliant cleaning action of the electrocleaning process can be observed.

Sodium Carbonate

A simple electrocleaning solution can be made from 10 g L^{-1} of sodium carbonate. Used at room temperature and around 6V, heavy fizzing will remove rust, paints, oils and other soils.

5.40b.

Periodic reverse cleaning can be very effective as it takes advantage of both anodic and cathodic cleaning. It is mainly used for removing heavy oxide deposits and scales. The main factor to monitor when employing this type of cleaning is the polarity of the last clean. Some metals will need this to be anodic and others cathodic. The best way to tell which way is correct is to look at the metal; if it looks clean with no discolouration or if you get rapid gassing in the post-cleaning acid dip then this is usually correct. Also check when plated that there is no skip plating or staining.

Filtration

Filtration of both cleaning and acid tanks has been found to be beneficial and can increase the tank life by up to 50%. Simple batch filtration, using between 1 –10 micron filter media, is adequate.

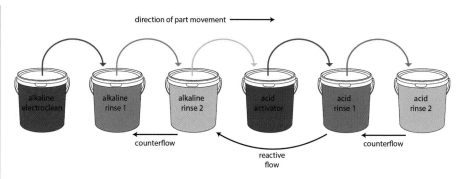

direction of part movement →

alkaline electroclean · alkaline rinse 1 · alkaline rinse 2 · acid activator · acid rinse 1 · acid rinse 2

counterflow · reactive flow · counterflow

5.41: A counterflow rinsing diagram.

RINSING

Rinsing parts after each cleaning stage is just as important as the rest of the cleaning processes. Rinsing with purified water, not simple tap water, removes excess solutions taken from the cleaning/plating tanks. This removal of solution stops contamination of the next stage or solution the part will enter. It also reduces staining or oxidization caused from small drips and spots of chemicals.

Purified water is the most used material in every electroplating and cleaning process, and the large amounts needed for each process can become very costly. Therefore, it is important to be efficient and effective with water usage, and there are many ways to achieve this. Many industrial plating plants have rinse flow systems that majorly reduce water consumption as well as reduce the amount of waste products. Two useful processes that are very popular in the electroplating industry are those of counterflow rinsing and reactive rinsing. As described by Fister [12],

counterflow rinsing is where 'the relatively clean rinse water from the second rinse in a rinse tank pair is flowed to the more contaminated primary rinse tank. Therefore, cleaner water is always moving to less clean rinse tanks. The cleanest water is still used for the critical final rinse, but the same rinse water is reused for the initial and least critical rinse'. It is good practice, to both conserve rinse water and to make it safe for disposal, to follow a counterflow rinse tank system. In a reactive rinse, the water from an acid rinse tank is used to replenish the water from an alkaline rinse tank. The acidity of the water can neutralize the alkalinity of the other tank. This reduces the drag-in contaminants for the next tank and conserves water.

As parts are removed from the tank they are often left to drip above it to allow for the materials to return. This can occasionally cause stains, marks and oxidization. An easy way to reduce these effects and to minimize the water quantities used is to spray the items as soon as they are pulled out of the cleaning/plating tank. The amount of water used to spray is a tiny proportion of an

5.42: Having a hand-held squirt bottle and applying this to items is effective in many ways. Firstly, it is quick and easy to use. A few sprays will release enough water to remove most of the excess solution.

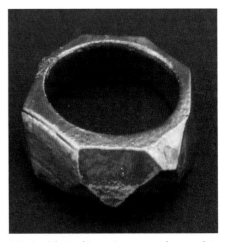

5.43: A typical rinse tank and cleaning tank system, starting with alkaline cleaner on the right, followed by a rinse, acid cleaner, and final cleaning rinse. The addition of the spray bottle helps reduce drag-in and drag-out contamination.

5.44: A quick visual inspection may not be enough to spot any remaining surface soils. This ring looks clean enough to go onto the next stage of plating.

extra rinse tank. It can also be done over the cleaning/plating tank itself with the run-off rinse water returning the material lost through drag-out into the tank and replenishing any water lost through evaporation. This small step can save a lot of money on water and materials and reduces the amount of potentially hazardous material that has to be disposed of. The downside to using a spray rinse is that some parts have complex geometries – the spray may not be able to reach each surface.

A rinse tank is therefore critical in removing the final parts of surplus solution. This tank will have minimal amounts of contaminants due to the spray pre-rinse, extending its life and reducing costs. Each cleaning and plating process should have its own rinse tank. For example, a system that contains a heavy-duty soak cleaner, an acid pickle, zinc plating solutions and a conversion coating will need a minimum of four rinse tanks. An effective way to reduce the number of tanks is to double dip. The process of removal and immersion removes more solution than simply leaving it in the rinse tank.

In most cases, warm water rinses are most effective, however, some porous metals such as some grades of cast iron and steel benefit from a series of cold and hot water rinses. All ferrous metals are better finished off with a cold water rinse. The lower temperature reduces the rate of evaporation and thus reduces the formation of oxides (rust).

TESTS FOR THE CLEANING PROCESS

It is very important to test the cleaning process from start to finish to ensure that the item or items are clean and ready for plating. These tests are easy to perform, they are:

◆ Visual examination
◆ Water break test
◆ White cloth test

With visual examination, you are looking for a clean surface without the slightest trace of soils or deposits on the items. If the cleaning process is borderline or insufficient it should be visually obvious. This is a straightforward thing to do and involves using a magnifying glass or some sort of visual magnification to scan the metal surface looking

5.45: Having a closer look, and looking through a magnifying glass, reveals that the surface is not quite fully clean, there are areas of oxidization and staining that have not yet been fully removed. If not carefully inspected, these faults may be missed and cause problems in plating in further processes.

for changes in appearance and colour. A visual inspection should always be done, however, the eye does not spot all possible problems, such as some metallic films that are indiscernible and will not be dissolved in the acid activation process. This may appear to plate well but will fail over time.

The water break test is, again, an easy test to carry out and is a marker for any surface grease or oil. Simply wet the

5.46: *All that is needed for a water break test set-up is a container of water, cloth for drying and the item.*

5.47: *This metal surface has not passed the water break test. As you can see, water is beading on the surface and does not create a continuous sheet.*

5.48: *You can see that most of this part has passed the water break test as there is a uniform film of water across much of the surface. The edges, however, show that there are still traces of soils that need to be removed.*

cleaned part. There should be a continuous film of water over the surface of the metal with no breaks in that film. If you have a break in the film, or you can see the water starts to bead into droplets, then this is an indication that there are hydrophobic soils on the surface. Examples of hydrophobic soils are: oils and grease, wax, polishing compounds and silicone. The water break test is normally performed after the final cleaning step and before acid activation. It should be carried out using distilled or deionized water. It can be done after acid activation, but items may fail at this stage and still be clean and activated for plating. If the acid activator contains a wetting agent, then it may pass a water break test but not be clean enough for plating.

A white cloth test is another quick and easy test designed to show the presence of inorganic and some organic soils that may be left after cleaning. It can be carried out at any stage of the cleaning process but is mostly used after the final cleaning step and after acid activation. The white cloth is wiped over the item and then carefully inspected for any black or grey smuts, oil staining or any other visual stain that can be seen on the cloth.

It the items fail any of these tests, then it is important to check the cleaners and activators for any possible contamination or problem with the cleaning process. Possible problems can be with the cleaners themselves or with the operating conditions such as temperature, time in the tanks, rinsing or time from tank to tank.

5.49: The white cloth shows that there are soils on the surface of the item and that it needs more cleaning before it can be plated.

COMMON FAILURE MODES ASSOCIATED WITH IMPROPER CLEANING

This section will list some of the common failures in electroplated parts that can be attributed to improper and insufficient cleaning.

5.50: After wiping this recently cleaned bolt on the cloth, no marks or staining appeared, meaning that it has passed the white cloth test and is ready to move onto the next stage.

◆ Blisters, Peeling, Delamination, Poor Adhesion:
If any of the above occur it can be an indication that oil and grease have not been completely removed from the part. It may also be caused by the incorrect polarity when using an electrocleaner or that the current is too high.

5.52: Peeling of zinc plating.

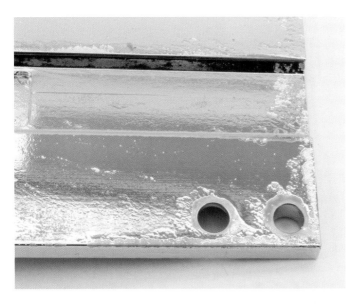

5.51: Blistering of nickel plating.

◆ Pitting:
If the pitting is localized to a small area, it can again be caused by the lack of removal of oil and grease. If pitting is widespread over the whole part it may indicate a problem with the plating tank.

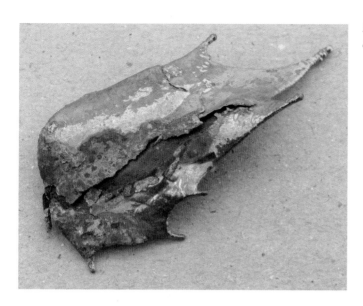

5.53: *Delamination of copper plating.*

5.55: *Stains due to contamination and poor cleaning of zinc plating.*

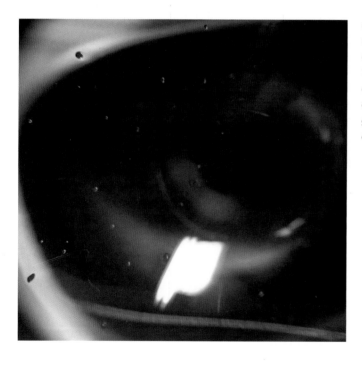

5.54: *This image shows micro-pitting under a 15× magnification. This can be caused either by tiny traces of soil or trapped gas from a too high current.*

◆ Roughness:
Failure to remove smut or solid particles will result in roughness along the surface. If using a vapour chamber, this can be a common problem as oils are removed but not solids. High-pressure solvent spray can be a solution for this. Incomplete rinsing in an alkaline tank or electro-cleaners may also cause roughness.

5.56: *Roughness of zinc plating.*

◆ Stains:
Visual stains are typically a sign that cleaning residue has been left on the part. This will be caused by poor rinsing of the part or by a delay in the rinse so that the water of the cleaning solution evaporates, leaving a film of chemicals. To combat this problem, you can lower the temperature of the cleaning solutions and decrease the transfer time between cleaning and rinsing. This problem may also be an indicator of incomplete soil removal, particularly oils.

6   The Electroplating Process

Pulling together all the previous chapters, we can begin to look at electroplating as a whole process with specific steps and stages. It can be split into several steps:

1. Clean
2. Rinse
3. Activate
4. Rinse
5. Electroplate
6. Rinse
7. Post-plating process

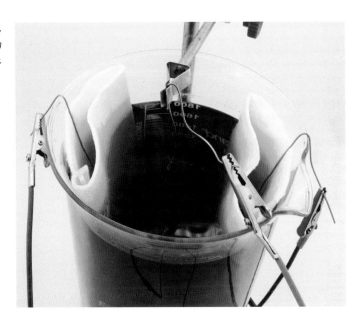

6.1: A typical, small, 1 litre nickel plating system for jewellery.

This first step, cleaning, is the most important one as it influences the rest of the process. While cleaning your parts may seem like a quick job, spending the extra time and patience to make sure that the surface is completely ready to take a new layer will greatly increase the quality of your results. There will probably be numerous cleaning steps you will have to follow, so make sure to rinse your part between each one followed by a final rinse at the end. Activation, while it comes under cleaning, is the last process that needs to be completed prior to electrodeposition. It is important to activate the surface to ensure strong adhesion between the base metal and the newly formed surface. Again, rinsing after this is important as it will reduce drag-out losses and minimize drag-in contaminants. Once

activated and rinsed, electroplating can begin. This is the most exciting part as the change in material and appearance can almost be instant. The operating conditions and electrolyte make-ups are in previous sections; this chapter will deal with specific common solutions and how to plate them onto common metals. Problems can occur, and the common failure modes will be covered in the next chapter. Once complete, rinse again before adding any final post-plating processes.

Test it First

If you are new to the electroplating process or are unsure on how your electrolyte will perform, do a small test with a spare part. This will ensure that you have the correct operating conditions and that your solutions give the required finish.

USING FARADAY'S LAWS TO ESTIMATE THICKNESS AND TIME

Before you begin, it can be advantageous to perform a little mathematics to work out how long you will need the part in the tank or how thick your new layer will be.

As mentioned in Chapter 1, Faraday's laws of electrolysis can be employed to work out the amount of material deposited on the part to be plated in grams. Taking nickel as an example, the amount deposited is directly proportional to the current and time. Starting from Faraday's first law:

$$m = \left(\frac{Q}{F}\right)\left(\frac{M}{z}\right)$$

For simplicity we will assume that during electrolysis the current is constant, so the total electric charge, Q, can be written as $Q = alt$, which is current multiplied by time multiplied by the efficiency of the current. This changes the equation to:

$$m = \left(\frac{alt}{F}\right)\left(\frac{M}{z}\right) = alt\,\frac{M}{Fz}$$

For nickel, $M = 58.6934$ g mol^{-1}, $z = 2$ and the Faraday constant is $F = 96485.33289$ C mol^{-1}, which is equivalent to 26.799 ampere-hours per mole. Putting those values into the equation gives:

$$m = 1.095 \times a_c It$$

The value of 1.095 is known as the proportionality constant for nickel. Other common metals have the pro-

portionality constants of: copper 1.186, zinc 1.220, gold 7.339, silver 4.025.

To get an accurate measurement for m, the current efficiency ratio or the electrode efficiency ratio must be found out. At the anode, the efficiency for nickel dissolution is practically 100%, so $a_a = 1$. The cathode efficiency is quite different. The efficiency may vary, depending upon the plating solution, to be around 90–97% efficient, meaning a_c can range from 0.9 to 0.97. This loss in efficiency is due to the hydrogen reaction at the cathode where some of the current is consumed.

Using Faraday's equation above, and extending our nickel deposition example, an expression for the average thickness of a nickel plate can be found. The thickness is the mass divided by the density multiplied by the volume. That is:

$$s = \frac{m}{dA} = 1.095 \times \frac{a_c It}{dA}$$

The mass of nickel deposited can be replaced by the previous equation $m = 100 \times 1.095 \times 0.95 \times It$ (multiplied by 100 to get thickness in microns), so the density of nickel is $d = 8.907$ g cm^{-2}. The equation becomes:

$$s = 11.680 \times \frac{It}{A}$$

We also know that $\frac{I}{A}$ is equivalent to current density, which for decorative nickel is around 4 A dm^{-2}, meaning that if we plate this part for two hours, the thickness will be:

$$s = 11.680 \times 4 \times 2 = 93.44\,\mu m$$

We can rearrange the equation to give us an estimate for time. Making time the subject gives the equation:

$$t = \frac{sdA}{1.095a_c I}$$

Inserting the values for a thickness of 10 microns would give:

$$t = \frac{10 \times 8.907}{100 \times 1.095 \times 0.95 \times 4}$$

$$= 0.214\ hours \cong 13\ minutes$$

The above equations only work for the average thickness over the surface of an item. The actual thickness in a certain area will vary depending on the current density in the area. This in turn depends on a few factors: geometry, location in tank and dimensions of tank.

BEGINNING THE ELECTROPLATING PROCESS

At this point you should have all your tanks, cleaning, rinse and electroplating in position, connected to a power supply and at the correct temperature. You should also have on the right PPE and be following the correct safety precaution. First, follow the cleaning methods. Once you have cleaned the work thoroughly, it is time to put them in the tank.

When putting the work in tank, make sure the power is off. A central location relative to anodes but close to the surface of the electrolyte is the best position; this placement will give the best overall current density to the parts to be plated.

6.2: A basic plating system with the workflow moving from right to left, starting with an alkaline cleaner, rinse, acid cleaner, rinse, nickel plating, and final rinse.

6.4b: Tank with bus bar.

6.3: A diagram of cathode placement. The work should be placed as centrally as possible within the tank, meaning the distance between the work and each anode should be the same. This will give a more uniform current density across the surface of the work, giving a more even metal finish.

Bus Bar

It can be useful, if not plating with a rack or barrel, to have a bar that spans across the tank to support the item. This bar is commonly referred to as a bus bar. It acts as a support to hang one or multiple items from ready to be plated. It should be made from copper and be cleaned after each use, allowing for optimal conductivity and ease of connection.

CHECKING DURING ELECTROPLATING

It may be tempting, but do not lift parts out of the tank to check them while plating, as this will cause a few problems. Primarily, burns and scorches will appear on the parts due to the decreasing area in the tank while the current remains constant – overcurrenting occurs causing burnout. Taking parts out in the middle of the process also promotes rapid oxidization of the newly formed atoms. This is especially problematic when using electrolytes with high temperatures. As you lift parts out, the solution evaporates rapidly, exposing oxygen to the surface. When you put the part back in the tank, the formation of surface oxides will become detrimental to the rest of the deposited metal. Delamination and blistering will occur due to poor adhesion between base metal, oxide and new layer. More information on failures can be found in the next chapter.

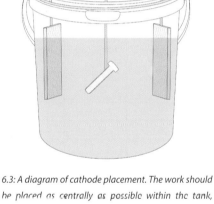

6.4a: This copper bar spans the top of the tank, above the electrolyte, to allow for the easy placement of parts. Copper works best due to its conductivity.

6.5: Burns caused by lifting the work from the tank while the current is on.

6.6: Zinc-plated steel fastenings.

If you need to check the part, first turn off the power. This will ensure no current-related problems occur. Then lift the part out of the tank, rinse and observe. If you are happy, reactivate the part in the same acid activator as before and continue to plate.

ZINC

Acid chloride zinc is one of the most common types of metal plating solutions used in small workshops. This is due to its ease of use, cost-effectiveness, corrosion protection and final finish. Below are the steps for plating of acid zinc onto mild steel.

Zinc-Nickel Alloy

The zinc plating processes outlined will be the same for zinc-nickel alloy plating.

Zinc		Deposited onto	Steel		
		Operating Conditions			
Steps	Process	Formulation	Temperature (°C)	Current Density (A dm^{-2})	Time (minutes)
1	Alkaline soak clean	Heavy-duty alkaline	70		1–5
2	Rinse	Purified water	18		
3	Acid soak clean	10–15% HCL (or other)	20–30		1–5
4	Rinse	Purified water	18		
5	Alkaline anodic electroclean	Heavy-duty alkaline	80	5–10	1–5
6	Double rinse	Purified water	18		
7	Acid activation	10–15% HCL (or other)	25		1
8	Rinse	Purified water	18		
9	Zinc plate	Acid chloride	27	1–2	10–20
10	Rinse	Purified water	18		
11	Acid activation	1% nitric acid solution	18		1
12	Passivation	Chromate conversion	18		0.25–0.5
13	Double rinse	Purified water	18		
14	Dry		< 20		

Table 6.1: Zinc plating process.

When plating high-carbon steels or high-tensile steel (P.C. >10.9 or 1000 N mm^2) then a de-embrittlement bake is necessary after step 10. This can be done in an oven at 200–220°C for four hours. Once complete, follow on from step 11.

The process outlined above will be the same for all types of zinc plating solution when covering steel. As you go through the process multiple times, there will be certain steps that you may want to change or tweak, such as the level of brighteners in the solution to give a more lustrous finish.

Zinc can also be plated directly onto copper, brass and bronze. The only difference with the process on these metals is the type of cleaners used. A light duty alkaline should be used for step 1 and a weaker acid activator solution for step 7. This can be sulphuric acid, used at a rate of 10 to 15% by weight.

NICKEL

Nickel is one of the most frequently deposited metals in electroplating. As a result, there are many solutions available to plate, the most common of which have been included in this book. Each solution has a different purpose; some are used as strike plates, some are barrier plates, some are used to increase corrosion resistance, some are used to activate the surface of the metal while others are used for their mechanical or decorative properties. This large variety of solutions gives rise to slightly different methods of deposition, and the most common are detailed below.

Nickel		Deposited onto	Steel		
		Operating Conditions			
Steps	Process	Formulation	Temperature (°C)	Current Density A dm^{-2}	Time (minutes)
1	Alkaline soak clean	Heavy-duty alkaline	70		1–5
2	Rinse	Purified water	18		
3	Acid soak clean	10–15% HCL (or other)	20–30		1–5
4	Rinse	Purified water	18		
5	Alkaline anodic electroclean	Heavy-duty alkaline	80	5–10	1–5
6	Double rinse	Purified water	18		
7	Acid activation	10–15% HCL (or other)	25		1
8	Rinse	Purified water	18		
9	Nickel plate	Any	As per Chapter 3		
10	Rinse	Purified water	18		
11	Dry				

Table 6.2: Nickel plating process.

Nickel		Deposited onto	Copper		
		Operating Conditions			
Steps	Process	Formulation	Temperature (°C)	Current Density (A dm^{-2})	Time (minutes)
1	Alkaline soak clean	Light-duty alkaline	70		1–5
2	Rinse	Purified water	18		
3	Acid activation	5–10% H$_2$SO$_4$ (or other)	20–30		0.5
4	Rinse	Purified water	18		
5	Nickel plate	Any	As per Chapter 3		
6	Rinse	Purified water	18		
7	Dry				

Table 6.3: Nickel onto copper.

Nickel onto Steel

The process for plating nickel directly onto mild steel follows the same steps as zinc in the previous section, until step 9. At this stage a Watts nickel solution should be used. Afterwards, rinse and dry; there is no need for passivation.

Nickel onto Copper

Nickel plating onto copper and copper alloys is common. Copper is used as both a filler plate to cover imperfections and as a barrier layer in multilayer processes. The stages for plating nickel onto copper differ slightly to that of nickel on steel.

Nickel onto Aluminium

Nickel plating onto aluminium requires the use of a zincate as one of the pre-plating processes. A zincate applies a thin layer of zinc on the aluminium while simultaneously removing any oxidization from the surface. Zincates have been covered in the previous chapters, so only an outline of the process and

Nickel		Deposited onto	Aluminium		
		Operating Conditions			
Steps	Process	Formulation	Temperature (°C)	Current Density (A dm^{-2})	Time (minutes)
1	Alkaline soak clean	Light-duty alkaline	70		1–5
2	Rinse	Purified water	18		
3	Aluminium etch	100 gL^{-1} sodium hydroxide 5 gL^{-1} sodium gluconate	18		0.5
4	Smut removal	30% nitric acid	18		until clean
5	Rinse	Purified water	18		
6	Zincate (immersion)	Proprietary solution	As per instruction included with solution		
7	Rinse	Purified water	18		
8	Zincate (surface removal)	Proprietary solution	As per instruction included with solution		
9	Rinse	Purified water	18		
10	Zincate (uniform colour)	Proprietary solution	As per instruction included with solution		
11	Rinse	Purified water	18		
12	Nickel strike plate	High sulphate	25		1
13	Rinse	Purified water	18		
14	Nickel plate	Any	As per Chapter 3		
15	Rinse	Purified water	18		
16	Dry				

Table 6.4: Nickel onto aluminium.

its application to aluminium will be covered.

Nickel onto Zinc, Tin and their Alloys

Nickel plating onto zinc and zinc alloys can be tricky. Many industrial platers do not usually plate onto zinc, it is therefore very useful to do the process at home. As with all electroplating, the crucial factor is to get the base metal prepared correctly. With cast zinc, very little cleaning needs to be done; only soak cleaning in a light duty alkaline cleaner followed by activation. The standard process for the plating onto zinc is a cyanide copper layer followed by nickel, however, we do not advise using cyanide-based plating systems, so an alternative is the use of a high-sulphate nickel strike plate. The same process is used to plate on tin and tin alloys.

Nickel onto Stainless Steel

Strike plating using a Wood's nickel strike solution is the most effective way to plate stainless and prepare it for following plates. The chrome content of

---

Acid Copper

At step 14, nickel can be replaced with high-throw acid copper to fill imperfections and improve corrosion resistance. If acid copper is plated at this step, follow instructions for nickel onto copper for the final nickel layer.

---

stainless steel makes it difficult for other metals to be deposited on top with good levels of adhesion. Most stainless steel grades contain between 10–20% chrome. Poor adhesion is down to the formation of a chrome oxide layer when exposed to oxygen. This quality of stainless steel is what makes it so popular as the oxide layer is excellent at corrosion and wear resistance because it can self-passivate. It is also difficult to remove chemically and mechanically. Using a high-chloride nickel strike can remove the chrome oxide and deposit a very thin layer of nickel in one go. In a similar way to the anodic cleaning action, hydrogen forms on the stainless steel surface during electrolysis and 'scrubs' the chrome oxide, lifting it from the surface and allowing nickel to be deposited. On 405 stainless a sulphamate nickel strike can be used instead of a Wood's nickel strike.

BLACK NICKEL

Black nickel plating is for decorative finishes. It is mostly plated over the top of mirror-finished bright nickel layers. It

Nickel		Deposited onto	Zinc/Tin		
		Operating Conditions			
Steps	Process	Formulation	Temperature (°C)	Current Density (A dm^{-2})	Time (minutes)
1	Alkaline anodic electroclean	Light-duty alkaline	70	2–3.5	0.5–1
2	Hot rinse	Purified water	50–80		
3	Double cold rinse	Purified water	18		
4	Acid activation	.05% H_2SO_4 (or other)	20–45		0.5
5	Hot rinse	Purified water	50–80		
6	Double cold rinse	Purified water	18		
7	Nickel strike plate	High sulphate	25		
8	Rinse	Purified water	18		
9	Nickel plate	Any	As per Chapter 3		
10	Rinse	Purified water	18		
11	Dry				

Table 6.5: Nickel onto zinc or tin.

Nickel		Deposited onto	Stainless Steel		
		Operating Conditions			
Steps	Process	Formulation	Temperature (°C)	Current Density (A dm^{-2})	Time (minutes)
1	Alkaline anodic electroclean	Heavy-duty alkaline	80	3.2	2
2	Rinse	Purified water	18		
3	Acid cathodic activation	6% H_2SO_4, HCL or other	18	10	3
4	Rinse	Purified water	18		
5	Nickel strike plate	Wood's: 240g nickel chloride 200ml 22% HCL	18	3.2	3
6	Rinse	Purified water	18		
7	Nickel plate	Any	As per Chapter 3		
8	Rinse	Purified water	18		
9	Dry				

Table 6.6: Nickel onto stainless steel.

6.7a: The regular finish of black nickel can be quite light if only a thin layer is applied.

6.7b: Anodic darkening of this nickel-tin alloy can deepen the colour, producing a rich black without altering the dimensions of the part. The part is made the anode of the circuit and the surface is intentionally oxidized to create the dark colour.

Anodic Darkening

If plating black nickel using a nickel-tin electrolyte, anodic darkening can be employed. This can be carried out using the same process as anodic activation. The length of time anodic darkening takes depends on the final colour needed. If darkened for too long, the newly formed black nickel layer will be dissolved.

can also be plated directly onto copper, brass, tin, pewter and zinc. The process for plating black nickel onto nickel is detailed below. For the deposition onto other metals, follow the process for regular nickel, simply replacing the electrolyte with black nickel.

Black Nickel		Deposited onto	Nickel (or other)		
		Operating Conditions			
Steps	Process	Formulation	Temperature (°C)	Current Density (A dm^{-2})	Time (minutes)
1	Alkaline anodic electroclean	Heavy-duty alkaline	80	3.2	2
2	Rinse	Purified water	18		
3	Acid cathodic activation	6% H_2SO_4, HCL or other	18	10	3
4	Rinse	Purified water	18		
5	Black nickel plate	As per Chapter 3			
6	Rinse	Purified water	18		
7	Dry				

Table 6.7: Black nickel onto nickel.

Pyrophosphate Copper		Deposited onto	Steel		
		Operating Conditions			
Steps	Process	Formulation	Temperature (°C)	Current Density (A dm^{-2})	Time (minutes)
1	Alkaline soak clean	Heavy duty alkaline	70		1–5
2	Rinse	Purified water	18		
3	Acid soak clean	10–15% HCL (or other)	20–30		1–5
4	Rinse	Purified water	18		
5	Alkaline anodic electroclean	Heavy-duty alkaline	80	5–10	1–5
6	Double rinse	Purified water	18		
7	Acid activation	10–15% HCL (or other)	25		1
8	Rinse	Purified water	18		
9	Pyrophosphate copper strike	As per Chapter 3	25	0.6–1.5	10
10	Rinse	Purified water	18		
11	Other metal plate				

Table 6.8: Pyro copper onto steel.

COPPER

Copper is a very versatile material to electroplate with and due to the various copper electrolytes, it is possible to copper plate onto most metals. Most of the common copper processes will be covered below.

Pyrophosphate Copper onto Steel

Pyrophosphate copper strike plating process onto steel.

Pyrophosphate Copper onto Zinc

Strike plating onto zinc, zinc alloys, zinc die-cast, tin, pewter and other tin alloys can be achieved with pyrophosphate copper. However, it must be carried out with a strike electrolyte; an electrolyte containing fewer copper pyrophosphate salts.

Pyrophosphate Copper onto Aluminium

Copper strike plating onto aluminium can be achieved using pyrophosphate copper.

Pyrophosphate Copper		Deposited onto	Zinc/Tin		
		Operating Conditions			
Steps	Process	Formulation	Temperature (°C)	Current Density (A dm^{-2})	Time (minutes)
1	Alkaline anodic electroclean	Light-duty alkaline	70	2–3.5	0.5–1
2	Hot rinse	Purified water	50–80		
3	Double cold rinse	Purified water	18		
4	Acid activation	0.5% H$_2$SO$_4$ (or other)	20–45		0.5
5	Hot rinse	Purified water	50–80		
6	Double cold rinse	Purified water	18		
7	Pyrophosphate copper strike	As per Chapter 3	25		
8	Rinse	Purified water	18		
9	Copper plate	Any	As per Chapter 3		
10	Rinse	Purified water	18		
11	Dry				

Table 6.9 (above): Pyro onto zinc and Table 6.10 (below) Pyro onto aluminium.

Pyrophosphate Copper		Deposited onto	Aluminium		
		Operating Conditions			
Steps	Process	Formulation	Temperature (°C)	Current Density (A dm^{-2})	Time (minutes)
1	Alkaline soak clean	Light-duty alkaline	70		1–5
2	Rinse	Purified water	18		
3	Aluminium etch	100 gL^{-1} sodium hydroxide 5 gL^{-1} sodium gluconate	18		0.5
4	Smut removal	30% nitric acid	18		until clean
5	Rinse	Purified water	18		
6	Zincate (immersion)	Proprietary solution	As per instruction included with solution		
7	Rinse	Purified water	18		
8	Zincate (surface removal)	Proprietary solution	As per instruction included with solution		
9	Rinse	Purified water	18		
10	Zincate (uniform colour)	Proprietary solution	As per instruction included with solution		
11	Rinse	Purified water	18		
12	Pyrophosphate copper strike	As per Chapter 3	25		1
13	Rinse	Purified water	18		
14	Copper plate	Any	As per Chapter 3		
15	Rinse	Purified water	18		
16	Dry				

Copper		Deposited onto	Nickel		
		Operating Conditions			
Steps	Process	Formulation	Temperature (°C)	Current Density (A dm^{-2})	Time (minutes)
1	Alkaline soak clean	Heavy-duty alkaline	70		1–5
2	Rinse	Purified water	18		
3	Acid cathodic activation	6% H_2SO_4, HCL or other	18	10	3
4	Rinse	Purified water	18		
5	Acid copper plate	As per Chapter 3			
6	Rinse	Purified water	18		
7	Dry or next plate				

Table 6.11: Copper onto nickel.

Copper		Deposited onto	Steel		
		Operating Conditions			
Steps	Process	Formulation	Temperature (°C)	Current Density (A dm^{-2})	Time (minutes)
1	Alkaline soak clean	Heavy-duty alkaline	70		1–5
2	Rinse	Purified water	18		
3	Acid soak clean	10–15% HCL (or other)	20–30		1–5
4	Rinse	Purified water	18		
5	Alkaline anodic electroclean	Heavy-duty alkaline	80	5–10	1–5
6	Double rinse	Purified water	18		
7	Acid activation	10–15% HCL (or other)	25		1
8	Rinse	Purified water	18		
9	Alkaline copper strike plate	As per Chapter 3			
10	Neutralization	5% H_2SO_4			5 seconds
11	Rinse	Purified water	18		
12	Copper plate	Any	As per Chapter 3		
13	Hot rinse	Purified water	18		
14	Cold rinse	Purified water	18		
15	Dry				

Table 6.12: Copper onto steel.

Copper onto Nickel

Copper plating onto nickel is often performed when nickel is used as a strike or to fill in imperfections in the nickel. Acid copper is ideal for this process due to its high throwing power and ability to fill in holes and scratches.

Copper onto Steel

The process of acid copper on top of an alkaline/pyrophosphate copper strike is one of the most common when plating on a steel substrate.

GOLD

Much of the total usage of gold plating is split between jewellery manufacture and electronics manufacture. Gold can be plated onto several different metals, however, when plated onto copper or silver, atoms from the base metal will diffuse into the gold causing staining. Gold is most often deposited onto nickel, stainless steel and white bronze.

Gold onto Nickel

See Table 6.13.

Gold onto Stainless Steel

Some types of gold solution will plate directly onto most grades of stainless steel. If you have any problems when plating onto a grade of stainless then activate and plate with a Wood's nickel strike before then gold plating.

Gold		Deposited onto	Nickel		
		Operating Conditions			
Steps	Process	Formulation	Temperature (°C)	Current Density (A dm^{-2})	Time (minutes)
1	Alkaline soak clean	Heavy-duty alkaline	70		1–5
2	Rinse	Purified water	18		
3	Acid cathodic activation	6% H_2SO_4, or other	18	10	3
4	Rinse	Purified water	18		
5	Gold plate	As per Chapter 3			
6	Rinse	Purified water	18		
7	Dry				

Table 6.13: Gold onto nickel.

Gold		Deposited onto	Stainless Steel		
		Operating Conditions			
Steps	Process	Formulation	Temperature (°C)	Current Density (A dm^{-2})	Time (minutes)
1	Alkaline anodic electro-clean	Heavy-duty alkaline	80	3.2	2
2	Rinse	Purified water	18		
3	Acid cathodic activation	6% H_2SO_4, HCL or other	18	10	3
4	Rinse	Purified water	18		
5	Gold plate	As per Chapter 3			
6	Rinse	Purified water	18		
7	Dry				

Table 6.14: Gold onto stainless steel.

Gold onto White Bronze

White bronze has now become a common barrier plate for on jewellery, taking over from nickel due to its allergenic properties.

SILVER

Silver plating occasionally has two stages, a silver strike followed by a thicker silver layer. Silver is more noble than most other metals, meaning when they are placed into a solver solution, electroless plating will begin to occur. This is the reason a silver strike is needed. The silver strike solution has a much lower concentration of silver and contains no brighteners, levellers or other additives. This combination lowers the rate of immersion plating, so the strike can apply a thin, well-adhered silver layer. In most cases, this strike layer is not needed.

Like gold, silver can also suffer from diffusion when plated over copper, brass and bronze. This only occurs with thin plates, less than a few microns thick.

Silver onto Copper and Copper Alloys

See Table 6.16.

Gold		Deposited onto	White Bronze		
		Operating Conditions			
Steps	Process	Formulation	Temperature (°C)	Current Density (A dm^{-2})	Time (minutes)
1	Alkaline anodic electroclean	Light-duty alkaline	70	2–3.5	0.5–1
2	Hot rinse	Purified water	50–80		
3	Double cold rinse	Purified water	18		
4	Acid activation	1% H_2SO_4 (or other)	20–45		0.25–0.75
5	Hot rinse	Purified water	50–80		
6	Double cold rinse	Purified water	18		
7	Gold plate	As per Chapter 3			
8	Rinse	Purified water	18		
9	Dry				

Table 6.15: Gold onto white bronze.

Silver		Deposited onto	Copper and Copper Alloys		
		Operating Conditions			
Steps	Process	Formulation	Temperature (°C)	Current Density (A dm^{-2})	Time (minutes)
1	Alkaline anodic electroclean	Light-duty alkaline	70	2–3.5	0.5–1
2	Rinse	Purified water	18		
3	Acid activation	5-10% H_2SO_4 (or other)	20–45		0.5
4	Rinse	Purified water	18		
5	Heavy silver plate	As per Chapter 3			
6	Boiling rinse	Purified water	100		
7	Rinse	Purified water	18		
8	Dry				

Table 6.16: Silver onto copper.

Silver		Deposited onto	Nickel		
		Operating Conditions			
Steps	Process	Formulation	Temperature (°C)	Current Density (A dm^{-2})	Time (minutes)
1	Alkaline anodic electroclean	Heavy-duty alkaline	80	3.2	2
2	Rinse	Purified water	18		
3	Acid cathodic activation	6% H_2SO_4, HCL or other	18	10	3
4	Rinse	Purified water	18		
5	Silver plate	As per Chapter 3			
6	Rinse	Purified water	18		
7	Dry				

Table 6.17: Silver onto nickel.

Silver		Deposited onto	White Bronze		
		Operating Conditions			
Steps	Process	Formulation	Temperature (°C)	Current Density (A dm^{-2})	Time (minutes)
1	Alkaline anodic electroclean	Light-duty alkaline	70	2–3.5	0.5–1
2	Hot rinse	Purified water	50–80		
3	Double cold rinse	Purified water	18		
4	Acid activation	1% H_2SO_4 (or other)	20–45		0.25–0.75
5	Hot rinse	Purified water	50–80		
6	Double cold rinse	Purified water	18		
7	Silver plate	As per Chapter 3			
8	Rinse	Purified water	18		
9	Dry				

Table 6.18: Silver onto white bronze.

Silver onto Nickel

See Table 6.17.

Silver onto White Bronze

See Table 6.18.

Silver		Deposited onto	Silver Strike		
		Operating Conditions			
Steps	Process	Formulation	Temperature (°C)	Current Density (A dm^{-2})	Time (minutes)
1	Alkaline cathodic electroclean	Light duty alkaline	70	2	1–3
2	Rinse	Purified water	18		
3	Acid activation	2–4% nitric acid	18		1
4	Rinse	Purified water	18		
5	Silver plate	As per Chapter 3			
6	Rinse	Purified water	18		
7	Dry				

Table 6.19: Silver onto silver strike.

Brass/Bronze		Deposited onto	Steel		
		Operating Conditions			
Steps	Process	Formulation	Temperature (°C)	Current Density (A dm^{-2})	Time (minutes)
1	Alkaline soak clean	Heavy-duty alkaline	70		1–5
2	Rinse	Purified water	18		
3	Acid soak clean	10–15% HCL (or other)	20–30		1-May
4	Rinse	Purified water	18		
5	Alkaline anodic electroclean	Heavy-duty alkaline	80	5–10	1–5
6	Double rinse	Purified water	18		
7	Acid activation	10–15 HCL (or other)	25		1
8	Rinse	Purified water	18		
9	Brasss/Bronze plate	As per Chapter 3			
10	Rinse	Purified water	18		
11	Dry				

Table 6.20: Brass/bronze onto steel.

Silver onto Silver Strike

If silver is plated immediately onto silver strike then no other steps are needed except for a quick spray rinse before transferring to the silver plating tank. If there is an increased amount of time between the strike and subsequent plate, follow the instructions below.

BRASS AND BRONZE PLATING

Brass and bronze are mostly used as decorative finishes on top of nickel, steel, copper and zinc die castings.

Brass/Bronze onto Steel

See Table 6.20.

Brass/Bronze onto Nickel

See Table 6.21.

Brass/Bronze onto Copper

See Table 6.22.

Brass/Bronze		Deposited onto	Nickel		
		Operating Conditions			
Steps	Process	Formulation	Temperature (°C)	Current Density (A dm^{-2})	Time (minutes)
1	Alkaline anodic electroclean	Heavy-duty alkaline	80	3.2	2
2	Rinse	Purified water	18		
3	Acid cathodic activation	6% H_2SO_4, HCL or other	18	10	3
4	Rinse	Purified water	18		
5	Silver plate	As per Chapter 3			
6	Rinse	Purified water	18		
7	Dry				

Table 6.21: Brass/bronze onto nickel.

Brass/Bronze		Deposited onto	Copper and Copper Alloys		
		Operating Conditions			
Steps	Process	Formulation	Temperature (°C)	Current Density (A dm^{-2})	Time (minutes)
1	Alkaline anodic electroclean	Light-duty alkaline	70	2–3.5	0.5–1
2	Rinse	Purified water	18		
3	Acid activation	5–10% H_2SO_4 (or other)	20–45		0.5
4	Rinse	Purified water	18		
5	Brass/Bronze plate	As per Chapter 3			
6	Rinse	Purified water	18		
7	Dry				

Table 6.22: Brass/bronze onto copper.

Brass/Bronze onto Zinc

See Table 6.23.

IRON

Iron plating is mostly used to apply a hard layer to soft metals such as plating onto copper printing plates or on the tips of soldering irons. Due to the limited applications for iron plating, we will just cover the process of iron plating onto copper.

6.8: A common use of iron is that of printing plates. The addition of iron allows for longer-lasting plates at a lower cost than nickel or copper.

Brass/Bronze		Deposited onto	Zinc		
		Operating Conditions			
Steps	Process	Formulation	Temperature (°C)	Current Density (A dm^{-2})	Time (minutes)
1	Alkaline anodic electroclean	Light-duty alkaline	70	2–3.5	0.5–1
2	Hot rinse	Purified water	50–80		
3	Double cold rinse	Purified water	18		
4	Acid activation	1% H_2SO_4 (or other)	20–45		0.25–0.75
5	Hot rinse	Purified water	50–80		
6	Double cold rinse	Purified water	18		
7	High-sulphate nickel strike	As per Chapter 3			
8	Rinse	Purified water	18		
9	Brass/Bronze plate	As per Chapter 3			
10	Rinse		18		

Table 6.23: Brass/bronze onto zinc.

Iron		Deposited onto	Copper and Copper Alloys		
		Operating Conditions			
Steps	Process	Formulation	Temperature (°C)	Current Density (A dm^{-2})	Time (minutes)
1	Alkaline soak clean	Light-duty alkaline	70		1–5
2	Rinse	Purified water	18		
3	Acid activation	5–10% H_2SO_4 (or other)	20–45		0.5
4	Rinse	Purified water	18		
5	Iron plate	As per Chapter 3			
6	Rinse	Purified water	18		
7	Dry				

Table 6.24: Iron into copper.

Before plating with an iron solution, the salts must be put through a reduction treatment. Most ferrous [Fe(II)] salts contain a proportion of ferric ions. These need to be removed before plating as a high concentration will result in a brittle, stressed and pitted iron surface. This treatment can be achieved by mixing the solution and then adding acid to decrease the pH to around 0.5. Once this has been done, add wire wool or degreased iron filings and leave for twenty-four to forty-eight hours. This will draw out the Fe(III). An alternative method of ferric ion removal is by dummy plating. In both methods, the completion of the reduction treatment is indicated by the colour of the solution; it should become clear and green with no yellow tint. This solution will remain free from a large concentration of ferric ions through regular use. If a bath is not used for a time, then treatment may be needed. The anodes should also be removed if the solution is left idle for more than a day.

Below a pH of 3.5, the acid within the electrolyte will become depleted. This is due to the difference in efficiency between the cathode, which is 100%, and the anode, which is in the range 80–99%. Regular checks are therefore needed to maintain the pH. In higher pH ranges, the electrode efficiencies are virtually the same and so the solution does not need to be checked as regularly.

7 Common Problems

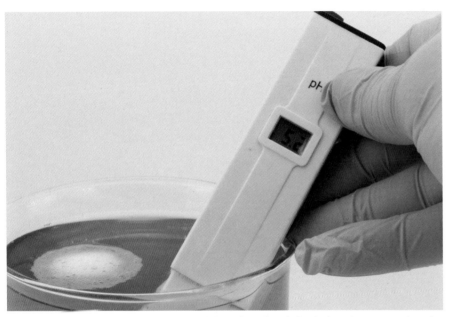

7.1: Plating system checks are necessary and should be performed regularly. Checking the pH is something that should be done before using the solution as drag-in and drag-out losses can cause changes. The regular operating of some solutions, nickel for example, will change the pH. Regular checks will notify you of these changes and allow you to adjust accordingly.

◆ Current density problems
◆ Additive imbalance problems (covers brighteners, Levellers and wetters, etc.)
◆ Filtration and anode bag problems
◆ Metallic impurities

CLEANING AND ITS IMPORTANCE

Cleaning is a critical process in electroplating; not only will it make the finish of the item better, but it drastically improves adhesion between the metal substrate and the newly deposited layer. Rinsing is a part of the cleaning and plating process that is important in the reduction of drag-in contaminants into plating tanks.

Metal parts that are newly made, or that have recently been cast, forged or machined, are easy to clean. They require minimal processes to achieve the right surface ready for plating. Older parts, in contrast, take more time and effort to prepare. The improper cleaning of older parts is usually the cause of cleaning-related problems in plating: poor adhesion, blistering, burns and drag-in contamination. Attention to detail is critical and examination of all surfaces is very important.

Monitoring, testing and evaluating the electroplating process are essential to ensure that results are consistent and that the plating solutions remain in optimum working condition. Solution maintenance is often overlooked until problems are seen in the finished items, and by then it is too late. You may have spent hours preparing, cleaning and plating only to find that the brightness is too low or that, due to a surface film, you get poor adhesion or blistering. Checking and monitoring temperature, pH, current density and agitation is critical to avoid problems. With a few simple checks, tests and routine maintenance these problems can be avoided, saving time, money and stress.

Here are the common problems that monitoring and testing will catch:

◆ Poor cleaning and subsequent drag-in
◆ Incorrect pH
◆ Incorrect temperature
◆ Poor agitation

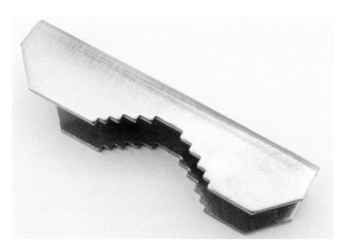

7.2: *A new part can be very clean and needs only a small amount of preparation to be plated on, leading to few problems.*

7.4: *The roughness and skip plating on this part are due to soils remaining on the surface. The part did not pass a water break test before being plated, likely due to contaminants in the rinse tank or oxides on the surface.*

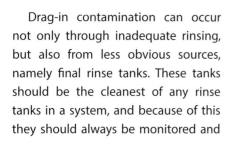

7.3: *Parts similar in quality to that of this image are the main cause of problems. High levels of corrosion can make cleaning difficult, which in turn can lead to failures of the final metal finish.*

7.5: *It can often be tricky to spot but iridescent sheens are evidence of surface films on rinse tanks.*

Drag-in contamination can occur not only through inadequate rinsing, but also from less obvious sources, namely final rinse tanks. These tanks should be the cleanest of any rinse tanks in a system, and because of this they should always be monitored and checked for any surface films and soils. Surface films from rinse tanks, due to drag-in contaminants, are not always obvious, showing no colour change or difference to the rest of the water. The films will wrap around an item as it is

Cleanliness Experiment

As an experiment for cleanliness of parts, imagine a pair of hollow steel rear foot rests from a motorcycle. They measure about 140mm long and are made from 2mm pressed steel with a spot-welded end plate and are open-ended at the mounting point. One footrest will be on the drive chain side of the motorbike while the other side is opposite the wheel. Cleaning and preparing the outside of the footrest is easy with either mechanical or chemical, or a combination of both. The results of cleaning on the outside are easy to see and to test, but the inside can't be seen. On the chain side, the hollow will slowly fill with grease and debris off the chain. On the opposite side, due to spray off the wheel, rusting will occur. The level of cleaning required to prepare the outside is not enough to clean the inside and the inside is often overlooked as no plating will occur here. As soon as the item goes into the plating tank the plating process itself, rather the formation of gases due to electrolysis, will start to release any remaining rust or grease. By simply plating a pair of steel foot rests you may have ruined your electrolyte! Any surfaces that can't be checked by the cleaning process tests should either be masked or sealed off.

7.6: Any film that is present on the surface of a rinse tank will be transferred onto the work when it is removed from the tank.

7.7: Drag-out from an acid sulphate copper solution can be seen in this rinse tank as it has a blue hue.

lifted out of the tank. In some cases, this film can cover the entire surface of an item. When the item is placed into the plating solution, the film will either continue to stick to the part or dissolve into the rest of the solution, causing a gradual increase in contamination.

The contaminants that cause a film in a rinse tank may also start to cause a film on the plating tank, leading to poor adhesion and skip plating depending on the severity of the film.

Another drag-in contamination source can be plating tanks and the transfer of parts from one plating solution into another; for example, transferring an item from a copper tank to a nickel tank. A simple immersion rinse after copper plating may not be enough to remove brightener film and plating solution, so more energetic

7.8: Dust and debris from mechanical cleaning.

7.10: Plating solutions should be clear and translucent, like this acid zinc electrolyte, and not milky or opaque.

methods (air agitation) may have to be employed.

Mechanical cleaning can be very effective, in relation to both the time and effort, at cleaning items. It can, however, be another source of drag-in contamination. Mechanical cleaning removes a layer of metal, oxides and soils from the surface of a part, forming dust and debris. This debris can become lodged in areas of the item, such as the threads of a bolt or nut. Pre-plating rinsing alone may not remove this before it goes into the plating tank. The orientation and positioning of the part may then allow this debris to become dislodged or it may remain in place and cause poor

adhesion and blistering of the subsequent plate. More vigorous rinsing and cleaning is required to remove this type of drag-in contamination.

CHECKING FOR DRAG-IN CONTAMINATION

The first step in identifying drag-in contamination is visual inspection. Any change in the appearance of the electrolyte, such as surface films, change in colour or cloudiness in the plating tank, can be an indicator of drag-in contamination and can easily be spotted. Oil films are easily identified on the top of a still, cool plating tank. Removing

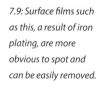

7.9: Surface films such as this, a result of iron plating, are more obvious to spot and can be easily removed.

a small sample of the electrolyte and placing it into a glass beaker will allow you to inspect the solution more easily. All plating electrolytes should be transparent. Some will be clear, some will be coloured (acid copper is blue and Watts nickel green), but they should not look cloudy or milky.

Ultimately, anything that goes into the plating tank can be a source of drag-in contamination, from the items you are plating to the heater you use or the filtration system. Evidence of drag-in contamination can be an unexpected change of pH. Poor rinsing after acid activation will lead to a slow lowering of the pH. Similarly, strong alkaline cleaners that are not properly neutralized by the acid activation process and subsequent rinse can lead to raising of the pH. Check the pH with a digital meter or litmus paper; if it is outside the normal range or expected value then there is contamination within.

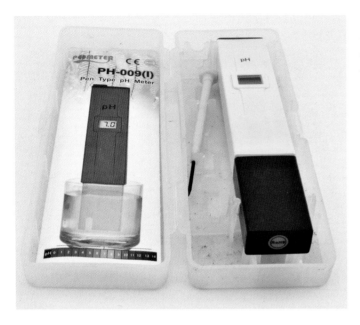

7.11: A digital pH meter is more accurate than litmus paper. Make sure to always clean it after use to give correct readings.

7.13: The dullness and dark marks on this washer after zinc plating form evidence of heavy metal co-deposition, contamination in the electrolyte and failure to remove some surface soils.

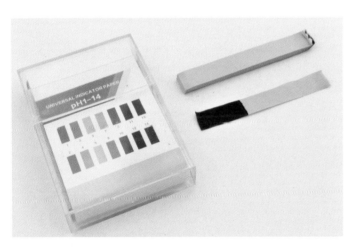

7.12: Litmus (pH) paper can be more difficult to read, especially if you are colour-blind. The scale provided only gives an estimation of the pH level, which is often not accurate enough for some plating solutions.

slowly and carefully. Make sure you are wearing PPE and have a container into which you can place the cloth and film residue. If this is done carefully, most if not all the film should have been removed. If not, then repeat this step, allowing the electrolyte to settle before dragging the cloth over the surface again.

If the electrolyte is cloudy or milky in appearance, turn off any agitation

One final way to monitor drag-in of heavy metals is through inspection of the newly formed surface layer. Spotting or dullness can be indicators of heavy metal contamination.

REMOVING DRAG-IN CONTAMINATION

Electrolyte surface films can be removed using polypropylene cloth, foam or felt. First, turn off any heating, agitation and filtration and wait for the electrolyte to cease moving. The lack of movement coupled with the cold temperature will cause the film to rise and become more obvious. Drag the cloth over the surface

7.14: Heavy surface films created due to drag-in contamination, about to be cleaned with cloth.

7.15a: The cloth is being dragged across the surface film, removing it from the solution.

7.16: Allowing a milky and contaminated electrolyte to settle in an undisturbed location will cause the larger, heavier contaminants to settle out.

7.15b: You can see that the surface film has been transferred onto the cloth and removed from the surface of the solution.

7.17: The contaminants will collect at the bottom of a tank to form a sludgy and fluffy precipitate. The remainder of the solution can then be siphoned off, leaving behind this sludge to be properly disposed of.

and allow it to settle. For some electrolytes, allowing it to stand for twenty-four to seventy-two hours will turn it clear as heavier sediments will settle to the bottom. Once settled, pump, syphon or carefully pour the electrolyte into a clean tank or container, leaving any remaining sediment. If this does remove the contaminants to make the solution clear, then repeat the process and also filter the electrolyte using very fine (5 µm) filter media.

Once you have removed any visual contamination, check and adjust the pH as necessary. Finally, perform a plating test to check for any irregularities or

7.18: A 3D-printed Hull cell used for testing electrolyte.

hollow areas. Inspect the final rinse tank after the acid activation step regularly. This is the final rinse before plating, so it must always be clean and free from surface films. Never go from one plating tank to another without thorough rinsing, activation and rinsing. Always clean any items such as heaters filters, pumps, thermometers, pH testers, etc, before they go into the plating tank.

7.19: Plating onto sheet steel can also show the levels of brighteners. This image shows low brighteners, giving a dull finish.

7.21: The holes in this part have been drilled and tapped, and have been sealed ready for cleaning and plating. Silicone sealer has been used as this will not corrode in the solutions and creates a good seal.

7.20: This image shows that the level of brighteners is correct, a shiny lustrous surface has been achieved.

INCORRECT PH

With all plating tanks there is an optimal pH range in which they operate. Keeping within this range is very important to the usage and results achieved from the solution. There are several plating solutions that will naturally increase in pH through regular usage; Watts nickel and acid zinc electrolytes are among these. This is due to the difference in efficiency of the anode and cathode, causing the formation of ions and changing the pH. Acid copper does not change much over time due to the anode and cathode efficiency being

problems with the solution and to check the level of additives used. This can be done either by performing a hull cell test [13], or regular plating of flat sheet steel. If dull or grey, add more brighteners into the solution and test again.

There are several ways to stop drag-in contamination from poor cleaning or rinsing. Firstly, always carry out the cleaning process checks that are listed in Chapter 5. Check any recessed or hollow items and if in doubt seal any

almost the same, essentially 100%. All plating baths should be checked on a regular basis, with nickel and acid zinc being checked daily when in use.

Symptoms of Incorrect PH

The effects of incorrect pH are not so obvious as the symptoms will not show unless they are further outside the normal operating range. Here are some signs of incorrect pH to look out for.

Low pH will lead to cloudy deposits in low, medium and high current density areas. This is due to the reduced effectiveness and possible destruction of organic additives causing poor levelling. Plating will show as poor throwing power and slow deposit formation in low current density areas.

High pH will lead to an alteration of mechanical properties, namely reduced ductility that will cause brittleness. High pH can also be seen through 'burning' of high current density areas while the rest of the item exhibits dull deposits.

Prolonged use of both nickel and

7.23: Low pH can also lead to low throwing power. The areas of lower current density, in the centre of this bend, show little and dull plating.

7.24: Nickel hydroxide can be easily spotted as a milky green precipitate, similar to that of milky green tea.

7.22: Having a pH that is too low, or lower than regular operating conditions, can lead to the dull deposits that can be seen here.

acid zinc plating tanks with high pH will lead to the formation of nickel or zinc hydroxide. Nickel hydroxide is seen as a light green, fine powder. This may become apparent by the electrolyte increasing in opacity as the amount of

7.25: When it has been mixed in an agitated solution, zinc hydroxide will give the solution a cloudy, milky appearance.

Type of Electrolyte	Lowering the pH	Raising the pH	PH Range
Watts Nickel	10% sulphuric acid solution	Nickel carbonate	3.5–5
Neutral Nickel	10% sulphuric acid solution	Nickel carbonate	5.3–5.8
Wood's Nickel	25% hydrochloric acid solution	Nickel hydroxide	0.3–0.5
Acid Copper	34% sulphuric acid solution	Copper hydroxide	0.5–0.9
Pyrophosphate Copper	10% phosphoric acid solution	10% potassium hydroxide solution	8–9.5
Acid Zinc (ammonium-based)	10% hydrochloric acid solution	Ammonia	5.5–6
Acid Zinc (potassium-based)	10% hydrochloric acid solution	10% potassium hydroxide solution	4.8–5.3
Alkaline Zinc	10% hydrochloric acid solution	Sodium hydroxide	9–15
Alkaline Silver	10% nitric acid	40% potassium hydroxide solution	9.2–10

Table 7.1: Adjusting the pH.

7.26: Once allowed to settle, zinc hydroxide will drop to the bottom of the tank, forming a cloudy white substance.

nickel hydroxide builds. Eventually, it will look like milky green tea. If movement of the electrolyte ceases, the nickel hydroxide can settle, falling to the bottom of the tank to form a layer of light green powder. Zinc hydroxide can be seen as a white powder that, like nickel, will make the electrolyte appear cloudy. Zinc hydroxide can be converted to zinc chloride by the addition of ammonium chloride. The pH should then be adjusted back to within operating range. Nickel hydroxide can be converted back to nickel chloride by the addition of HCL.

Checking the PH

The most accurate way to check the pH is with a digital meter. These are readily available and are more accurate and cost-effective over the lifetime of the plating system when compared with disposable pH papers. Meters should be calibrated regularly to maintain accuracy. pH papers can be used but they will only give a rough guide that will be much less accurate than its digital counterpart.

Adjusting the PH

The pH can be adjusted by adding specific substances into the solution, as per Table 7.1.

TEMPERATURE CONTROL

It is important to control the temperature for each type of plating electrolyte. Temperature mainly affects the efficiency of the plating process, but it will also affect how brighteners and other additives, such as wetters, work. Extremes of temperature outside the regular operating range will cause these additives to stop functioning, leading to dull deposits with nodules. Another effect of temperature is how the metal lattice is deposited on the surface of the base metal. Lower temperatures can lead to wide micro cracking, decreasing the tensile strength of the new surface layer. With alloy plating, such as brass, temperature can affect the deposition rate of each metal, altering the properties and appearance. When brass plating, an increase in temperature is coupled with an increase in copper deposition rates, and this will lead to a red brass plate.

7.27: *Nodules and dullness are caused by high temperatures stopping the brighteners/levellers working. This is an extreme example, where the area of highest current density, the tip, has not been levelled due to the high temperature.*

7.28: *Red brass due to increase in temperature in brass plating solution.*

CURRENT DENSITY PROBLEMS

Maintaining the correct current density while plating is extremely important. Not only does it control the rate of metal deposition, but it is also responsible for the uniformity of deposits and their structure and appearance. Current density is the measure of current per surface area of the total amount of items in a tank. It is therefore important to calculate the total surface area accurately so that the correct current can be applied. This not only ensures efficient plating but will stop associated low and high current density problems. Reducing the difference between high and low current density areas on a part will reduce related problems such as burning or treeing, brightness or hazing and thin plating. The control of current density is critical when items need to be plated to a specific thickness.

Additive Imbalance Problems

There are several different types of additives used in the plating tank and their use has been outlined in Chapter 1.

An imbalance of brighteners is a common problem within electroplating and is usually easily determined through appearance of the newly plated surface. A hull cell test [13] can be employed to show the drop in brighteners before it appears on an item. Alternatively, plating onto an L-shaped sheet or extrusion will show the same problem. If brighteners are low, both tests will show dullness in low current density areas. As the level of brighteners decreases, the amount of dullness will increase, spreading to

POOR AGITATION

Agitation is important for several reasons. It helps to increase plating efficiency through increased particle transport rates and the removal of inhibiting particles and compounds from the electrodes. Agitation of the electrolyte will help remove hydrogen bubbles from the surface of the items being plated. These bubbles can be visible on the final surface layer if not properly removed, showing up either as tiny pits or dots. Having sufficient agitation dislodges the bubbles as they form. Agitation also helps maintain homogeneity of the electrolyte, making the solution more uniform in both concentration and temperature. Temperature grading or stratification can lead to increased rates of gas pitting as well as uneven metal deposition. In plating solutions that use large amounts of brighteners, such as zinc plating, agitation ensures that there is a constant supply of brightener at the metal surface. If this does not occur, the lack of brighteners will lead to a dull plate. A further effect of poor agitation is increased burning in high current density areas. This will be visible as dark staining on edges or areas close to the anodes. Increasing or ensuring the correct amount of agitation will help reduce this.

areas of medium current density. At this stage, it will become noticeable on regular items. Some brighteners are used more quickly than others. Acid zinc for example, uses brighteners quickly compared to bright nickel.

Another brightener problem prevalent in acid zinc plating is called 'brightener oil out'. As water evaporates from the electrolyte, the chloride ion concentration increases, leading to oversaturation of the solution. This forces the dissolved brighteners to precipitate and form an oily deposit on the tank and everything inside. The easy way to avoid this is to top up the electrolyte when water begins to evaporate.

In electroplating, wetters decrease the surface tension, which aids plating in two ways. The first is to stop evolved hydrogen causing pitting on the surface of the plate. This works by reducing the surface tension and allows agitation in the electrolyte to release the tiny bubbles in the electrolyte. The second is to aid run-off, which leads to few losses by drag-out of electrolyte. If the electrolyte runs low in wetters then pitting and drag-out losses will occur.

FILTRATION AND ANODE BAG PROBLEMS

All types of tank plating solutions benefit from filtration. Filtration removes insoluble particles that have been brought into the electrolyte by drag-in, use of impure chemicals, corrosion of equipment or anodes and airborne debris. Having an efficient filtration system will maintain both the quality of the plating results and the cleanliness of the electrolyte, leading

7.29a: The results of the Hull cell test show a low brightener level. The area of highest current density, the left side, is highly burnt and the rest of the plate is dull.

7.29b: Addition of more brighteners and levellers reveals a large difference in the Hull cell test results. The high current density area shows much less burning and the rest of the plate looks uniform and shiny.

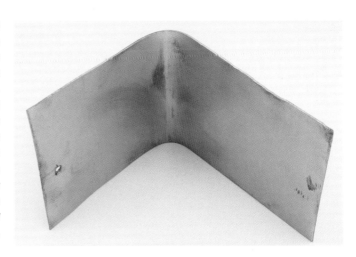

7.30: L-shaped sheet steel can be used as an alternative test to the Hull cell test to low brightener levels. The inside of the bend will have the lowest current density and the outer edges will have a higher current density. This variation will show the extent to which the brighteners and levellers are working.

to a longer electrolyte life.

Spotting problems with filtration are straightforward. If the output of a pump-based system has slowed, the filter will be blocked and will require cleaning or replacing. With an induction system this may be more difficult to spot, so scheduled routine cleaning

7.31a: This internal filtration system has been removed from the electrolyte as the clarity of the solution was decreasing. Upon lifting the filter from the tank, dark soils began to drop from the filter, clearly showing that the filter material needed changing.

7.31b: Cleaning the filter in hot water revealed the amount of soils collected in the filter; the water turned a murky black.

7.32: Dirty electrolyte. Oddly enough, this level of contamination will still produce a bright zinc finish.

7.33: A collection of large copper particles and anode sludge has gathered in the bottom of this anode bag. If these had not been removed, the resulting copper deposits would have become dull and rough.

7.34: Often it is necessary to have two anode bags. You can see from this image that the smaller, inner bag has become dark, containing high levels of particles, smuts and soils. The larger outer bag will prevent leaching of the soils from the bag into the electrolyte. Eventually, both bags will need to be cleaned and replaced.

should be carried out. Look for visual signs on the plated item such as grain- or sand-like deposits on the horizontal surfaces. Changes in the clarity of the electrolyte can also be an indicator

7.35: Iron impurities will often show as black spots where it has been deposited and rapidly oxidized.

7.36: Silicone impurities will often lead to hazing of the surface.

Inspect your Bag

Periodically inspect anode bags for rips and tears. You should also inspect for clogging, which will be seen as lodged anode particles in the anode bag material. Any hole in the bag will allow for the transfer of unwanted contaminants.

---

of filtration problems. Alkaline zinc is prone to clouding and needs more filtration than most other types of electrolyte.

Anode bags should be used with all types of plating as they create an easy and cost-effective form of anode filtration. The fine filter media around the anode stops large metal particles and impurities entering the electrolyte. Instead, they are contained within the bag. When anode bags are not used these particles can be carried throughout the solution and kept in suspension by agitation, migrating toward the cathode and eventually becoming deposited in the forming layer. This leads to a sand-like finish or pits from dislodged particles.

Anode bags should be tight fitting. The reduced space will cause the build-up of dropped particles to touch the anode, thereby receiving current. The current will provide electrons to reduce the metal particles to ions, which can be transported through the filter media and deposited on the cathode.

Double bagging is also recom-mended in most cases. Anode bags can be made of different materials depending on the type of electrolyte they are used for. Most common are polypropylene, nylon, cotton, viscose and polyester. Anode bags are rated on either thread count or on micron size. The higher the thread count, the smaller the particles it will stop. With micron size it is the lower number that relates to higher thread count. Some anode bags also have a polythene layer on the inside or outside at the bottom of the bag to act as a sludge trap.

Metallic Impurities

Evidence of heavy metal build-up in the electrolyte will first be observed on the plated item. Metallic impurities are often brought in to the plating tank by the corrosion of unplated parts, incorrect wiring, anode impurities or from parts that have accidentally fallen into the plating tank and have not been removed. Metallic impurities can also be introduced through drag-in from impure chemical additions. Here are some of the effects seen from common metallic impurities in a bright nickel plating tank.

7.37: *Using a 15× magnification, the needle deposits caused by calcium can be seen on this part.*

7.38: *Almost all metals that should not be in a solution will cause black deposits when codeposited. The darkness in this leaf is caused by iron stripped from a steel anode.*

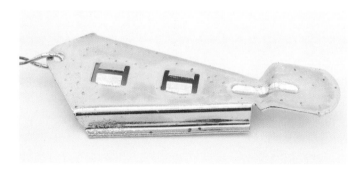

7.39: *Often, metal contamination may not show until a post-plating process is applied. Here, iron particles have reacted and oxidized with the chromate conversion coating to leave dark spots.*

Iron: Dissolved iron can cause hazing or pitting and reduce ductility. Iron may also precipitate to form ferric hydroxide, which can cause fine roughness on the plated part.

Aluminium and silicon: Can cause hazing in high current density areas or very fine roughness.

Calcium: Can build up in the plating tank when using hard water; this can cause fine needle-like roughness. This is due to the precipitation of calcium sulphate that occurs when the saturation point of $0.5gL^{-1}$ at 60°C is exceeded.

Copper, cadmium, lead, tin and zinc: These metals can deposit in low current density areas. This may cause hazing and dark or black deposits.

The most common contaminant for acid zinc electrolytes is iron. Iron contamination does not show in regular plating but in the post-plating process of chromate conversion coating. Dark spots or patches will become visible in the high current density areas. Iron can be removed using hydrogen peroxide, which will oxidize the Fe^{2+} to Fe^{3+}. The Fe^{3+} precipitates to form a sludge, which settles at the bottom of the tank. This sludge is also effective at capturing and removing other impurities. The precipitate must be removed from the solution either by filtration or by decanting. Failure to remove the Fe^{3+} sludge will result in reduction back to Fe^{2+}, causing the same problems to be repeated.

There are two common ways of removing metallic impurities; one is by using low current density electrolysis or 'plating out'. The other is to use high pH treatments (with acidic plating electrolytes). 'Plating out' is done by plating onto a relatively large piece of cleaned and prepared scrap metal at current densities well below the normal operating range. This process should take between twelve to twenty-four hours. Using this method, several metallic impurities are removed, including copper, zinc, lead and tin. By plating at current densities around 0.2–.05 A dm^{-2}, the minimum amount of anode material will be dissolved and used. All the other normal plating parameters apply, and using the correct tank temperature, agitation and filtration is still important. It is good practice to perform a 'plate-out' on a regular basis ensure the levels

7.40: *This is the result from a heavy metal plate-out. Visual inspection of the plated out scrap metal will reveal dark deposits, nodules and excess growth over most of the surface, and especially in low current density areas. This is evidence of heavy metal contamination. Plating out these metals with a low current is one of the easiest ways of removing this type of contamination.*

7.41: *Bleed-out of trapped acid copper electrolyte has caused the corrosion of copper and the growth of copper sulphate crystals. Liquid became trapped between the layers of this 3D-printed dinosaur skull due to insufficient application of sealant and conductive paint.*

of heavy metals remain low.

Using high pH treatments, iron, aluminium and silicon impurities will precipitate into their basic complexes, metal hydroxides, at a pH of 5.0–5.5. These can then be removed by regular filtration. For nickel, the electrolyte is heated to 65°C and a slurry of nickel carbonate is added until the required pH is achieved. The electrolyte is then allowed to settle for ten to eighteen hours before filtering the electrolyte.

Another option other than removal is to use complexing or sequestering agents to complex or sequester the unwanted metals. Several organic products based on citrates, gluconates or benzene sulphates are available as proprietary products.

ELECTROLYTE BLEED-OUT

Under some circumstances, electrolyte can become incorporated and trapped under the newly deposited metal surface. Over time, the liquid will diffuse through the metal and evaporate into the air, leaving behind and material in the electrolyte, metal salts. This often happens in electroforming applications where a porous item has not been sealed properly and has absorbed some electrolyte; when metal is deposited at very high rates; or when the electroplated layer is very thin. Resolutions for bleed-out are the proper sealing of porous objects, reduced plating speed and either thicker deposits or the addition of a dense nickel strike layer.

8   Post-Plating

8.1: Great-looking, well-plated parts with a variety of post-plating finishes. (Dave Cherry, DC Sandblasting)

Recently plated parts look great! They are shiny, clean and spotless. However, as soon as the parts reach the air, oxidization will immediately begin to happen, leading to corrosion. Unless used for art or demonstrations, corrosion is a very undesirable process. Post-plating treatments are critical in protecting a recently plated part from corrosion and are vital in creating long-lasting, oxidization-resistant parts.

When it comes to post-plating treatments, the field of surface engineering overlaps and plays an important part in deciding what finish and treatment is best for you. Surface engineering is the multidisciplinary field that tailors the properties of a surface or near-surface regions of a material to improve form and function. Electroplating is already a surface engineering technique, but the newly created metal plate will also need treatments to help with desirable properties. These properties include:

- Corrosion resistance
- Oxidation resistance
- Wear resistance
- Increased toughness or tensile strength

Surface Treatment/Coating Type		Treatment Benefits
Changing the metallurgy	Surface hardening (flame, laser, induction)	Improved wear resistance
	Shot peening	Improved fatigue strength
Changing the chemistry	Phosphate conversion coating	Corrosion resistance, plating/painting adhesion, lubricity
	Chromate conversion coating	Corrosion resistance, improved adhesion for paints or other organic coatings, providing decorative finish
	Black oxide conversion coating	Decorative (blueing)
	Anodising	Corrosion resistance, improved aesthetics, ability to take dyes, increased paint adhesion
Adding a surface layer	Electroplating/electroless plating	Corrosion resistance, wear resistance, electrical properties, appearance
	Paints	Corrosion resistance, wear resistance, electrical properties, appearance
	Dyes	Aesthetic appearance
	Lacquers	Corrosion resistance, wear resistance, electrical properties, appearance
	Oils	Corrosion resistance
	Thermal spraying	Wear resistance, corrosion resistance, oxidization resistance
	Vapour deposition (chemical and physical [CVD, PVD])	Improved wear, corrosion and erosion resistance, optical and electronical properties

Table 8.1: Metal surface treatments.

◆ Electronic resistance
◆ Aesthetical appearance

On the previous page is a table of a range of surface engineering techniques that can be applied to all metals and some electroplated parts. There are three main ways of altering the surface: changing the metallurgy, changing the chemistry or adding a surface layer or coating. The options highlighted are those that are most commonly used as post-plating treatments and will be discussed in this chapter.

The main treatment type is often determined by many factors: component size and accessibility, the corrosive environment, the anticipated temperatures, component distortion, the coating thickness attainable and cost.

All the methods above work to create a surface coating that controls corrosion. This is done by making the surface passive, or by adding a barrier layer.

PASSIVATION AND CONVERSION COATINGS

Passivation is one of the most common forms of post-plating processes. It refers to the process by which a metal surface becomes more passive. That is, it becomes less reactive with the environment to maintain a high level of corrosion resistance or for aesthetic purposes. Passivation is the spontaneous formation of a non-reactive micro-coating on the surface of a metal that is created in a chemical reaction with various materials, the most common of which is air. This passive outer coating is usually only a few nanometres thick

and is often an oxidized form of the base metal, or a chemically added metal, that inhibits further corrosion.

8.2: Post-plating processes will add an extra layer of protection onto parts. This image shows a part that has been passivated with a short immersion in a chromate conversion coating, adding a very thin layer of chrome and an iridescent shine.

Conversion coatings are a method of surface passivation involving a chemical or electro-chemical process used for corrosion protection, colouring or as paint primers. A few of the main examples of conversion coatings are:

◆ Chromate conversion coating
◆ Phosphate conversion coating
◆ Black oxide
◆ Anodizing

Phosphate conversion coatings, black oxidizing and anodising can be performed on electroplated surfaces but these are very uncommon. The most common conversion coating is chromate.

Seven Steps to Coating

If a conversion coating is not added immediately after electroplating, then the typical steps involved in adding a conversion coating are very similar to those of plating, that is: 1) cleaning, 2) rinsing, 3) acid etch/acid pickle, 4) rinsing, 5) conversion coating, 6) rinsing, 7) drying.

Chromate Conversion Coatings

A chromate solution is used to make the surface of a newly electroplated part passive by chemically altering the surface, converting it to a different material and structure. Solutions are grouped by their main active element, the derivative of the group's name, chromium (Cr). Chromate conversion coatings increase the corrosion resistance of plated parts significantly, and can also be used to alter the colour. The two common colours are blue and yellow, but there are different ones available including browns, greens and blacks. Chromates are widely used to finish aluminium, zinc, steel, magnesium, cadmium, copper, tin, nickel, silver and alloys of the metals mentioned.

Immersion or spraying are the two main ways of applying a chromate. However, there are other methods of application including brushing, roll coating, dip and squeegee, electrostatic spraying and anodic deposition in special cases. The safest way is the

8.3: Parts that have been immersed in a clear chromate conversion coating gain a protective barrier but show little change except for a shift in hue.

8.4: An example of the colour that can be achieved using a yellow-coloured chromate conversion coating.

8.5: Various chromate conversion coatings. From left to right; olive drab, yellow/gold, clear and black.

immersion method as it has the least risk of contact with the skin or inhalation. This process involves submersing the part into a bath of the chromate solution for a short amount of time –

8.6: The process of applying a passivate is simple – immerse for a short time and then rinse.

depending on the finish required – followed by an immediate and thorough rinse to remove all traces of the solution. It is also known as the Cronak process.

The Cronak Process

The Cronak process, developed and patented by the New Jersey Zinc Company in 1936, is the most widely used chromate conversion process. It involves the submersion of a zinc (alloy) part or plated part in a chromate solution for ten to fifteen seconds. The solution has two main components, sodium dichromate and sulphuric acid.

The chromate conversion solution acts in two ways. The first is to etch the surface of the metal, the second is to re-passivate the surface of the exposed metal by forming complex chromium and metal compounds. The chromium then rapidly oxidizes and forms chromium oxide, a barrier layer that delays the oxidization of the base metal or metal plate. The chrome that is not oxidized acts to re-passivate the metal whenever the surface is slightly worn by oxidizing, healing the passive layer. The chromium contained in conversion coatings comes in two valences: trivalent and hexavalent.

Trivalent Chromate

Trivalent chromates, $Cr(III)$, are a relatively new development in passivation, driven by the demand for materials to be hexavalent chrome-free.

Hexavalent Chromate

Hexavalent chromate, $Cr(VI)$, is the more traditional passivation solution used in automotive manufacture. It is now not as popular due to chronic health and environmental concerns.

In large-scale processes, chromate conversion coatings have come under pressure due to their effect on the environment, leading to new Restriction of Hazardous Substances (RohS) legislation.

Chromium-Free Passivation

Currently there are few chromium-free conversion coatings, and these few coatings only act on a limited number of metals: aluminium, zinc and steel. These conversion coatings are based on zirconium, silicates or titanium-

8.7a: *Statue of Alfred, Lord Tennyson, at Lincoln Cathedral showing the patina bronze collects when left to the open environment.*

8.7b: *The name plate of the statue shows even more clearly the blueish patina that occurs naturally on metals over time.*

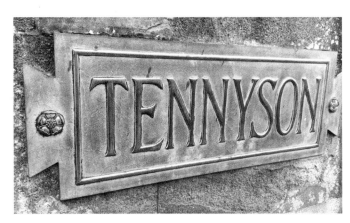

containing materials. The performance of chromate conversion coatings still cannot be matched by alternatives, so their use is limited to certain food industries and is not widespread. There is a move to develop these coatings due to the increasing, justifiable pressure for a reduced environmental impact.

PATINATION

The word *Patina* originates from the word *patena*, which referred to a shiny black varnish applied to shoes. It was later defined as patina by Italian Filippo Baldinucci, and was widely used to describe the way in which the varnish in paintings darkened over time. As the field of archaeology began to develop, the word was used to describe the coloured corrosions formed by buried bronze artefacts. In terms of electroplating, patina and patina chemicals refer to the chemical reaction (oxidization) formed on the surface of a metal that usually gives the metal a different colour and the chemicals that are used to do so. Patinas can be applied to many metals but the most common are copper and copper alloys (used in jewellery and art) and steel (used in large sculptures and artworks). The range of colours achievable and the ease of finding the chemicals and mixing the formulas is why patinas are so widely used as a post-plating process.

8.9: *Copper patina using potassium chloride and ammonium chloride.*

8.10: *Copper patina using ammonium chloride and sodium chloride.*

8.8: *Copper patina using a copper sulphate solution and ammonium chloride.*

There are a number of ways in which patinas are applied to metals:

◆ Spraying: Spraying a patina solution allows for good coverage and makes a good first layer for further patinas to be applied. It should be noted that it can be difficult to control the spray so that it does not leave patches or streaks.

◆ Brushing: Brushing is especially useful when it comes to variation of application. The thickness of the bristles can play a major role in the final finish of the patina, especially if applied with a stippling action. Foam-tipped brushes can also be used for a more even and controlled application.

8.14: One of the simplest ways of applying patina is through immersion in a patina solution. Leaving it for an extended period will increase the intensity of the patina.

8.11: Application of a cold liquid patina solution using a spray bottle.

8.13: Stippling with a patina solution can again give unique and artistic results.

◆ Packing: Moistened wood shavings or rice are a common medium for this method of patination. The part to be patinated is packed in a container with the packing medium that has been moistened with the patina solution. The results are a mottled and stippled finish. As a guide, the moistened medium should not drip when squeezed. If the medium is too moist then patchiness may occur.

◆ Wiping: Wiping a solution-soaked cloth over the metal is another method of applying a patina. This first wipe is usually followed by a second wipe with a damper cloth and the metal is then left to dry.

◆ Fuming/Vapour: Exposure to fumes or vapours will colour a metal surface over time. Leaving a part in an air-tight container, above and away from the solution, will result in surface patination.

◆ Immersion: Immersing the metal in the solution can be a great way to achieve a uniform patina over the surface of an object. The colour intensity is achieved either by a long duration or multiple immersions in the solution.

8.15a: Packing an item in material soaked with a patina solution can lead to some very interesting results. Here rice has been soaked in a solution containing sodium hydroxide.

8.12: The effect of applying the patina solution with a brush can be quite striking and give rise to some very artistic patterns.

8.15b: After two hours the patina is beginning to show and the outlines of each grain of rice can be seen.

8.15c: After twenty-four hours, the patina has reached a great and deep blue colour.

intensity of the applied patina. The hot method is usually achieved by applying the patina solution, followed quickly by a torch or heat gun over the painted areas.

Surface Preparation for Patinas

Like cleaning before electroplating, proper preparation of the metal surface for a patina is paramount. If the patina is being applied immediately after plating (with a rinse of the item in clean water first) then there is no need to worry about surface preparation. If the part has been left for a while, then care and attention will be needed to get the part ready to take a patina.

The same preparation as before electroplating can be used to clean a metal for a patina; abrasive cleaning, soak cleaning, solvent cleaning, detergent spray or wash and electrocleaning. Follow the previous chapters to prepare the metal.

Wear Gloves

Make sure to always wear gloves as any chemicals and residues on your finger will affect the cleanliness of the metal and therefore the patina.

The first three methods are the most common and can be classed as hot or cold. In the cold process, the patina solutions are applied directly, by brushing or spraying, and allowed to dry naturally at room temperature. A benefit of the cold process is the ability to control the colour intensity and texture. The major downside is that each application of the patina solution needs sufficient time to dry and react with the surface, and this is usually measured in days. Hot patination is much faster and is the preference for most artists. In this method, the patina solution is made at a higher temperature and the metal is also kept at a higher temperature. The temperature can affect the colour

Patina Colours

There are hundreds of patina formulae for different metals, both commercially available and home-made. A quick internet search will bring up long lists of different solutions. Copper and its alloys are usually the most common metal to be patinated due to the range of colours copper salts can be. Below is a table of some of the most common colours and solutions.

8.16: Copper patina using cobalt sulphate in solution.

8.19: This patina has been achieved using heat and a naked flame.

8.20: Bronze patina through a quick immersion in an alkaline cleaning solution with no rinse.

Copper is Key to Colour

If you want to colour a metal for aesthetic purposes only and physical tolerances of resistance are of little consequence, then it is usually easier to apply a thin copper plate over the top to gain access to a wider range of colours.

8.17: Copper patina using sodium carbonate and sodium hydroxide.

There are thousands of patina recipes available, too many to fit in this book! Below are a few examples of some simple and complex solutions and the colours they give.

Post-Patina Protection

Once the patina has been applied and the required finish is achieved it is necessary to protect the surface to stop any further changes. Humidity, oxygen and other chemicals in the atmosphere will cause the colour to change further

8.18: Copper patina using ferric nitrate.

over time. Before applying a protective coating, it is important to remove any residual water or dampness as excess water can lead to the formation of chlorides and sulphates on the surface of the metal, which can lead to corrosion. To dry an item thoroughly, it can be left in a dry environment for a day or can be left in the oven at a low temperature for a few hours. If you are using the hot method of application, then parts will dry much quicker. It is still important, however, to check over any areas that may have moisture.

There are several ways to seal a patina; each method applies a barrier layer that inhibits the diffusion and reaction of oxygen. Lacquers and metal sealers are one way and will give a very durable finish, though some are prone to cracking and chipping. Another option is to use a wax-based product. Try to avoid animal-based waxes, such

8.21: Adding a lacquer to a patina can often intensify the colour as well as sealing the surface.

8.22: Adding a wax sealer is a great way to seal the patina, however, care must be taken when applying the sealer as it can remove some of the surface.

as beeswax, as methods of collecting it are detrimental to both the animals and the environment. Museum wax or Renaissance wax are effective and so are some of the silicone waxes. It may be beneficial to buff or polish a patinated surface. If doing so, the waxes and polishing compounds can act as a sealant, too. When applying the sealant, be careful of the patina. If applied cold, it can be delicate and be removed with the vigorous rubbing. If applied hot, then this is not a problem.

PAINTING

Despite the benefits and complexity of some metal-finishing treatments discussed, the most commonly used is paint. Paints have four basic components: resin, solvent, pigment and other miscellaneous materials. The resins, often called binders, are organic compounds such as acrylic, epoxy or vinyl. The solvent is usually water or another organic compound. Pigments are the important part as they provide the main functionality of the paint. They can act as rust inhibitors, decrease permeability of the paint, give colour, or increase erosion and UV resistance. Typical pigments include zinc-based compounds such as zinc phosphate, zinc molybdate, zinc phosphorus silicate and zinc chromate. The miscellaneous compounds act to increase the usability of the paint and can be things such as dryers and flow control compounds. The ingredients of paints vary depending on the conditions of their final use. Below is a table containing various environments, applications and paint resins to give you an idea of the types of paints needed for different parts and uses.

Surface Preparation

As with electroplating, the importance of surface preparation for painting cannot be overemphasized. Surface preparation will impact durability and corrosion resistance, so that even the best paint applied in the most skilled way will fall short of its maximum performance. Similar to electroplating, paints that are not bonded to the surface, due

to soils, can flake and delaminate. Once plating has been completed, the surface should be made passive through the methods mentioned previously before being painted.

Lacquers, Sealers

Lacquers and sealers come in many different forms.

8.23: This lacquered ring will remain bright and lustrous due to the lacquered surface layer. No oxidization will be able to occur as the surface is sealed.

Cellulose

Plus points: relatively inexpensive, hard wearing, good for items that will remain indoors, quick drying. Negative points: hazardous fumes when applying, can breakdown in with ultraviolet light and cause staining under the lacquer.

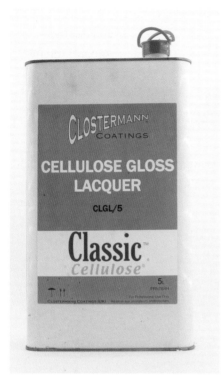

8.24: A standard cellulose lacquer, which can be used for sealing electroplated parts.

Polyurethane

Plus points: comes in both solvent and water-based forms, hard wearing, heat resistant, good against chemical attack, good against water ingress, easy to obtain, can be used on items for internal and external use.

Negative points: slow drying, hazardous.

Acrylic

Plus points: very affordable, quick drying, less toxic than cellulose or polyurethane, hard wearing, clear finish.

Negative points: runny consistency so tricky to apply evenly, can give a milky finish if applied to thickly.

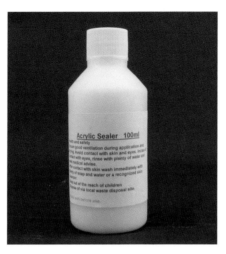

8.25: Acrylic sealer used for metals; and non-metals before plating.

Ceramic

Plus points, very hard wearing, can be applied as a very thin coat, very good water resistance.

Negative points: very expensive, can be hazardous to apply, may require baking.

8.26: Wax polish for furniture, antiques and woodwork can also be used as a metal sealer.

Wax

Plus points, leaves a natural looking finish, easy to apply by cloth, several different types.

Negative points: not hard wearing, multiple coats may be needed, time-consuming to achieve finish required.

Oils

Plus points: fast application (often dipped), relatively inexpensive.

Negative points: limited corrosion protection, not hard wearing, messy to apply.

8.27: This blend of oils works as a great sealant for metals, however, simple and common oils (such as WD40) can be used as a sealer.

When using a lacquer or sealer the part to be sealed should be clean and dry. Make sure you chose the right one for the type of metal you want to seal. Look for anti-haze and non-yellowing if you want to maintain a good appearance. When choosing a lacquer or sealer

8.28: Heat oxidization of metal causes different colours to be shown on the surface, from deeps reds to purples and blues.

The range of colours is almost limitless when it comes to dyeing in this method. However, organic pigments are susceptible to bleaching or discolouration if exposed to sunlight for extensive periods.

Colouring can also be achieved by heating the metals to promote oxide growth. When these oxide layers are formed, they incorporate gases from the atmosphere to give them different colours, ranging from pale golds and yellows to deep blues and blacks. Placing metal into boiling water can also change their colour, again, by promoting oxide growth. Certain media can be added to the water, such as salts or other metal oxides, to create specific colours. This is essentially creating a patina. These colours would only be for decorative purposes as a small amount of abrasion would remove the thin colour oxide layer.

HEAT TREATMENT

Heat treatment is used to reduce hydrogen embrittlement. During the cleaning and/or plating process, excess hydrogen may be introduced into the surface. This is because the cathode efficiency is not 100% and can happen due to excess heat, assisted by a high concentration gradient of hydrogen ions outside the metal. The hydrogen ions attach to the surface and begin to diffuse through the crystalline metal, slowly recombining in crystalline defects or vacancies to form hydrogen-based particles, therefore increasing the pressure within the metal. This pressure increase within the metal leads to a reduced ductility, toughness and

there are a few properties to consider. Will the part be for internal or external use? Will it be subject to abrasion or wear? Does the part flex, if so will the sealer flex or is it brittle? Do you want a natural look? Is corrosion protection more important than appearance? Is it easy to apply? Is it safe to apply?

DYES AND COLOURING

Colouring in metals usually occurs due to oxidization. The metal oxide either changes the colour of the metal, such as in patinas, or acts as a porous surface for organic dyes of inorganic pigments. Typically, the application of dyes is a dipping process. The part to be dyed will have recently been passivated (anodized, phosphated, or sulphated) and is then submerged into a solution containing the dye or pigment. The part is then removed and then sealed in boiling water. The colour is fixed in place due to the reduction in size of the pores due to heat and hydration.

tensile strength, leading to sporadic and unpredictable failures at stresses much less than those at which the metal is rated. This is hydrogen embrittlement. While this problem can affect various metals, such as steel, aluminium and titanium, it is high-strength steel that is of most importance. Applications for high-strength steel are mainly focused in the automotive industry, where the specific mechanical properties are needed. Under high stress loads, hydrogen-embrittled parts may fail catastrophically and unpredictably – which is not something that you want to happen in your car. The main way to treat hydrogen embrittlement is through heat treatment. The baking treatment reduces the amount of hydrogen within the metal and therefore reduces internal pressure. The best results come from prolonged heat treatments, up to twenty-four hours at 200°C. However, some metals may not benefit from a prolonged bake and would only need three to four hours

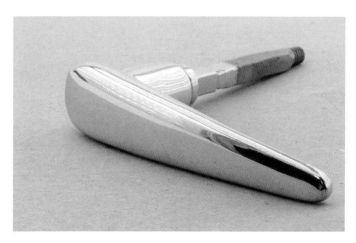

8.29: Sanding, buffing and polishing have all given this handle a mirror finish by first smoothing out any remaining pits and then levelling the surface mechanically.

at the same temperature or one to two hours at 300°C. This will liberate around 30% of the hydrogen and increase the time taken to fracture by an order of magnitude.

MECHANICAL TREATMENTS

Some mechanical methods, such as those mentioned in the cleaning section of this book, may be used once plating has been complete. For example, some badly pitted parts may still contain pits after plating. Here, buffing and polishing may be employed again to smooth down the region ready for another plate to fill in the remaining pits. Peening gives a unique aesthetical appearance and so can be used as a post-plating treatment, especially on jewellery items. Sandblasting may also be used if the desired finish is matte.

9 Safety

equipment. Performing a risk assessment will also help avoid any unnecessary risks and how to do this is described below. This chapter ends with further links in regard to safety and disposal of electroplating solutions.

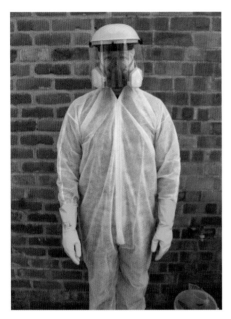

9.1: Full-body PPE is not always necessary for home plating but safety precautions should be taken at all times during the plating process.

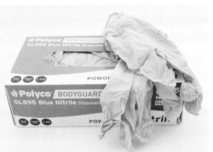

9.2: Disposable gloves, like these, should be used once and then disposed of. They are very resistant to corrosion but are very thin and can be split easily. For more protective handwear, use thick rubber gauntlets.

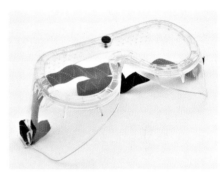

9.3: Goggles are essential when using agitation in solution or whenever movement of liquids is involved.

9.4: Dust masks and respirators should be worn at almost all stages of the electroplating process, even when there is adequate ventilation. There are various types available so check the requirements of your plating solution before buying one.

Operator safety is paramount in the entire electroplating process. Due to the nature of chemicals used and the potential for risks, care and attention must be employed when undertaking any part of the process. PPE will give you a protective barrier against fumes and vapours as well as any accidental leaks or spills. A description of the requirements of PPE is outlined later in this chapter. Understanding the safety symbols and phrases on material safety data sheets, MSDS, is a necessity when it comes to preparing and choosing

BEFORE PLATING

The most important health and safety advice is to wear PPE at every stage of the electroplating process. PPE includes protective eyewear – goggles, glasses, full-face masks – corrosion-resistant gloves, dust masks and respirators, aprons and body suits. It is also very beneficial for personal safety to practise good housekeeping. Keeping your plating area clean by removing any spills, shop dust or mess as soon

as it occurs will not only reduce health-related risks to accidental exposure but help minimize plating contaminants and increase the longevity of all your plating solutions.

Secondly, always refer to the material safety data sheets supplied with the

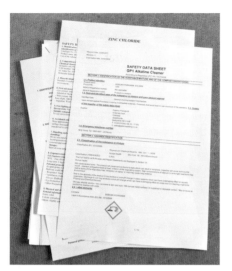

9.5: *Always read safety data sheets supplied with chemicals you buy. If you don't have any, request them.*

9.6: *Respirators are the best form of mouth, nose, throat and lung protection. They are relatively cheap and much more reliable than dust masks.*

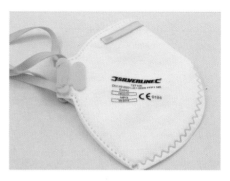

9.7: *You can find all the information on this FFP1 dust mask; its rating according to the different standards, certification, manufacture date and expiry date. Make sure to always check for these whenever you buy any PPE.*

9.8a: *An example of a carbon dust mask. The grey colour is an indication of the impregnated carbon.*

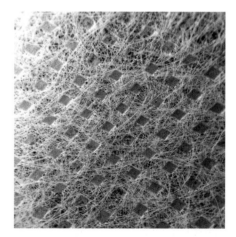

9.8b: *Magnified, you can see the carbon within the material of the dust mask. This layer of carbon adds extra protection over non-carbon masks.*

chemicals, also called MSDS. These will contain a comprehensive collection of relevant safety information, including correct usage, storage, instructions and pictograms and safety phrases.

Personal Protective Equipment

Personal protective equipment (PPE) is one of the most important aspects of staying safe. Completion of the risk assessment and awareness of the chemical hazards gives more information on the levels of PPE needed throughout the entire electroplating process. There are numerous pieces of equipment available and various ratings on each, so deciding exactly what you need can be difficult. When choosing equipment, check the MSDS for each chemical to see if a certain type is advised. If not then there will be an occupational exposure limit (OEL) that gives guidance on the rating of PPE. In practice, it is best to wear safety gear

that is rated higher than the chemicals you are using.

Wearing PPE should be expected when plating but in case you have a major accident make sure that there is plenty of water available to wash any splashes off the skin, face or eyes. Spills should be neutralized immediately or contained by use of an absorbent material and disposed of according to MSDS.

Dust Masks and Respirators

Dust masks and respirators are essential in situations where extraction is limited or when dust and vapours are in the immediate vicinity. Make sure that any respiratory protective equipment you wear has been tested and conforms to the European Standard EN149:2001 [14]. This standard dictates that all products must be tested to provide protection against solid and liquid aerosols. They are then given one of three classes:

◆ FFP1 will give up to four times OEL
◆ FFP2 will give up to twelve times OEL
◆ FFP3 will give up to fifty times OEL

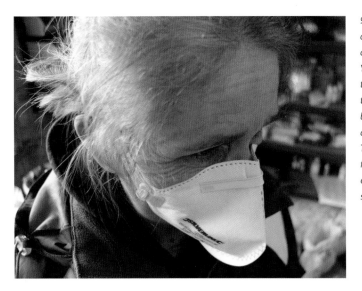

9.9: Make sure your dust mask forms correctly to your face. You can see here that there is a gap between the mask and the bridge of the nose, and at the side of the nose. These gaps will allow fumes and vapours to enter your respiratory system.

face shield, which can be inexpensive and offer complete face protection. If you feel uncomfortable wearing a face shield, then goggles with indirect ventilation are the next best option.

Gloves

Two types of gloves are recommended when mixing and plating, nitrile and natural rubber. Both are rated for use with the types of acid and alkaline solutions commonly used in electroplating. Gauntlets or longer gloves should be used to protect more of the body.

Check the Expiry Date

Dust masks will have expiry dates; remember to check these and note how long they can be used for.

9.10: Face shields, in conjunction with a dust mask or respirator, offer the best protection against spills, splashes, fumes, and vapours.

9.11: This type of glove will be ideal for all stages of the electroplating process. They are thick, which will reduce the chance of holes and tears, they are chemically resistant and will cover the wrist and part of the arm.

FFP1 has the lowest efficiency and is usually given to basic, disposable and single-use masks. These may be fine for the initial mixing but will be out performed by FFP2 and FFP3. FFP2 and 3 would be suitable for setting up the whole system and some extended usage. Masks that have been impregnated with activated carbon offer increased protection through chemical filtration and carbon adsorption. These are easily recognized as they appear greyish in colour. Respirators will have a higher safety rating than dust masks and would be suitable to use for the whole plating process.

It is important that your mask/respirator fits your face correctly. There are many that are supplied as 'one size fits all' but first try them on and make sure there are no gaps and that there is no air escaping or entering from the sides.

Face and Eye Protection

Face and eye protection can come in three forms: goggles, glasses and shields. The most effective option is a

Gloves will need replacing regularly. They will retain soils, which can be transferred into solutions or onto parts, and they will eventually corrode due to prolonged usage. When handling and cleaning parts ready for and after the plating process, both thin latex and nitrile gloves are suitable for use.

Body Protection

Disposable coveralls should be used when mixing chemicals in either powder or liquid form. Aprons may also be worn over the top of coveralls for added protection, depending

9.12: It is important to have a safe working environment. This workshop has a dedicated ventilated area that has been lined with PP plastic to reduce damage caused by any spills.

them for damage and make sure they are working correctly. This may be more difficult for glass tubular heaters as turning them on to check when they are not immersed in liquid will cause the glass to crack. Simply place in water three-quarters ¾ of the way up the shaft and check. Tears in chemical bags or cracks in chemical containers may lead to spillages and build-up of harmful gases in storage, These gases can become toxic or explosive, so make sure to check and store them in a ventilated area.

Setting up the System

Again, before doing anything make sure you are wearing PPE. This is a point that we will make regularly as it is vital. Read any safety data sheets you have and follow any advice they give. It is important to have a dedicated area in your workspace that is set up with correct extraction and adequate safety precautions like those in the image.

Placement of Tubs

Refer to Chapter 5 for the cleaning rinse flow diagram. These will advise you

on the solutions and processes being employed. If aprons and rubber boots are advised on the MSDS make sure that the apron is long enough and goes past the top of the boot to the top of the foot.

WHEN CAN PROBLEMS OCCUR?

Chemical exposure and safety risks can occur at almost all times in the electroplating process from receiving the chemicals and equipment, to the large number of post-plating treatments and even in the disposal of rinse water.

Pre-Plating and Pre-Cleaning Equipment and Storage Checks

Before you begin with any stage of electroplating, make sure you are wearing PPE.

It is vital to check each piece of equipment and chemical containers before you begin. Chemicals and com-

ponents are not always treated with the care and attention they should during transportation and this can lead to cracks, faults and tears.

Cracks in electroplating tanks may not be obvious initially but can develop when stresses, such as electrolyte pressure and heat, are applied. This can lead to leaks, spillages and catastrophic failures that can expose you to hazards. Equipment faults can lead to other hazards such as electrical shock. Before using each piece of equipment, check

9.13: During the initial set-up and mixing, or when moving chemicals, it is a good ideal to have disposable coveralls on. They will block most chemicals from contacting your skin or clothes.

9.14: Along with safety equipment and ventilation, having a clear and ordered space for electroplating will reduce the potentials hazards to yourself and the environment.

on where to place your tanks so that minimal dripping occurs. Make sure you place them in a well-ventilated area or that you have specific ventilation for the main tanks. Make sure the area is not busy – this will reduce the risk of accidental spills and reduce contamination of different areas.

Bunding

Selecting the correct, non-corrosive container has been outlined in a previous chapter and checking it for damage has already been mentioned. Something that has not been mentioned is the bunding of chemical tanks. Bunding, the placement of a secondary wall or container, is a legal requirement in many industries as it prevents further leakage of hazardous materials. When thinking about home plating, bunding your cleaning tanks, plating tanks and post-plating tanks can help minimize the risk of spillage if a tank splits or fails.

9.15: A bunded container greatly reduces the risk of spills and leaks.

Chemical Mixing

When mixing chemicals, make sure you add the chemicals to the tank first, then the water. This will reduce any unnecessary splashing or fizzing as the chemicals dissolve. If it is possible, it is safer to pump water, or any liquid, from one container to another, rather than to simply pour them. Pouring tends to create uncontrollable splashes and aeration, whereas pumping is more con-trolled and will reduce vaporization. Mix carefully and slowly. Once mixed and the solution is homogenous, you can begin to add other equipment.

Wiring the Tank

Wiring the tank, hanging the anodes and connecting them to the power supply, as well as installing other electrical equipment, poses electrical hazards, as mentioned previously. When setting up

9.16: *Chemicals should be added to the container first, followed by distilled, deionized or RO water. The chemicals in this image are those of Watts nickel; nickel sulphate, nickel chloride and boric acid.*

a tank, this can involve connecting the power supply or hanging the anodes, make sure the connections are secure and that there is a minimum amount of exposed wire. This will reduce the risk of shocks and shorts. Using home-made power supplies can be dangerous; they will not have passed industry safety standards designed to reduce harm. Make sure that you have minimized risk and that you are confident that everything is wired correctly.

Installing Equipment

Before installing any electrical components, make sure they are turned off and not plugged in to an electrical source. When you have performed quality checks for damage, turn on and test:

◆ Heating: If using a glass tubular heater, make sure it is immersed in solution before turning on as heating when not in solution can lead to cracks and breaks in the glass. With a warm solution comes fumes – you

can use fume control balls (croffles), which will cut out around 95% of mist and fumes and help to insulate the tank, reducing energy consumption.
◆ Filtration: Check that the filter is functioning correctly, that there are no blockages and that there is a consistent flow in and out.
◆ Agitation: Check that the agitation system is not causing too many fumes/bubbles, etc – change placement for safer plating if needed.

Cleaning Processes

During the cleaning stage there are a number of risk factors of which you should be aware. Firstly and most obviously, read the MSDS supplied with the chemicals. Take some time to assess the risk sources at each step of your cleaning stage. The most common are:

◆ Chemicals
◆ Fumes and exothermic heat from mixing
◆ Fumes and gases from electrocleaning
◆ Electrical shock from electrocleaning
◆ Solvents from solvent cleanings and degreasing
◆ Heavy metal dusts from grinding, buffing and polishing
◆ Spills and splashing from items transfer

Plating and Post-Plating Processes

The electroplating and post-finishing processes carry similar risks to the cleaning stage, fumes and vapours being the main risk source due to the increased corrosiveness and hazardousness of the electrolyte. Special care must be taken

when using chromate conversion coatings as these are especially harmful.

Chemical Storage

When storing chemicals and pieces of equipment there are a few general principles that should be followed:

◆ Store hazardous substances in a lockable, cool, enclosed but ventilated area.

9.17: *A chemical storage container, such as this, should be used to store the most hazardous of chemicals. It is secure, lockable and has ventilation holes so hazardous fumes are not created.*

◆ Store incompatible substances separately – cyanides and acids form an explosive mix – and avoid the risk of mixing and cross-contamination.
◆ Make sure all labels remain intact and visible.
◆ Limit access to the storage area.
◆ Ensure flammable, explosive or toxic materials are stored away from possible sources of electrical sparks, heat or flame.

Check all containers for leaks and spills – make sure lids are tight and secure.

9.18: *Always make sure labels are intact and readable. Failure to label your products may result in dangerous mixes of chemicals.*

Chemical Disposal

Waste from electroplating can have a tremendous effect on the environment. While home plating produces relatively small volumes of waste, their impact on the local environment is still just as detrimental. It is tempting to simply pour waste down the drain but this still puts potentially hazardous chemicals into the water system without correct treatment. Similarly, taking electroplating waste to the landfill can put hazardous materials into the environment if containers leak and let solutions flow into the ground.

The first step in waste disposal is avoidance of disposal. Trying to make your plating system more efficient to reduce the waste produced is the easiest step. The biggest source of waste in electroplating comes from drag-out – the loss of materials when parts are moved from one tank to another. Through proper practice, closed loops rinsing systems, multistage rinsing

and increased drainage time, drag-out losses will be reduced. Spent acid pickle and alkaline cleaner can be used to neutralize each other and reduce their environmental impact during disposal.

There are, however, times when you want to completely dispose of an electroplating solution, either because of major contamination and loss of function or because you no longer need it. In either case there are guidelines and regulations describing how this waste should be treated. These can be found in the next section.

To dispose of waste material correctly, place in a secure container and label the waste. The waste must be labelled with its contents and waste classification. This can be found in the Waste Classification guide [15] in the next section. Once labelled, it must be transported safely to a waste disposal centre. The transportation service used must be insured to carry hazardous waste, not just hazardous materials. A tool to find your nearest hazardous waste disposal service can be found in the next section.

Further Safety Information

Here are some sources of further information on safety regulations and disposal:

Health and Safety Executive (HSE)
'How to carry out a COSHH risk assessment' [16]
Online: www.hse.gov.uk/toolbox/harmful/coshh.htm

Environment Agency
Tel: 0870 8506506
Email: enquiries@environment-agency.gov.uk

Online: www.environment-agency.gov.uk
'How to comply with your environmental permit: Additional guidance for: The Surface Treatment of Metals and Plastics by Electrolytic and Chemical Processes (EPR 2.07)' [17]
Online: https://www.gov.uk/government/uploads/system/uploads/attachment_data/file/297005/geho0209bpip-e-e.pdf

Department for Environment Food & Rural Affairs (DEFRA)
Nobel House
17 Smith Square
London
SW1P 3JR
03459 33 55 77
https://www.gov.uk/guidance/contact-defra
'Waste legislation and regulations' [18]
Online: https://www.gov.uk/guidance/waste-legislation-and-regulations

GOV.UK
'Find a local hazardous waste disposal service' [19]
Online: https://www.gov.uk/hazardous-waste-disposal

Natural Resources Wales, SEPA, NIEA, Environment Agency
'Waste Classification: Guidance on the classification and assessment of waste – Technical Guidance WM3' [15]
Online: https://www.gov.uk/government/uploads/system/uploads/attachment_data/file/427077/LIT_10121.pdf

References

[1] S. La-Niece, *Metal Plating and Patination: Cultural, technical and historical developments*. Elsevier, 2013.

[2] L.B. Hunt, 'The early history of gold plating,' *Gold Bull.*, vol. 6, no. 1, pp. 16–27, 1973.

[3] E.H. Oakes, *A to Z of STS Scientists*. Infobase Publishing, 2014.

[4] G.R. Elkington, 'Improved process for gilding copper, brass,' US741 A, 17 May 1838.

[5] *Repertory of patent inventions and other discoveries and improvements in arts, manufactures and agriculture*. Macintosh, 1838.

[6] 'The Industrial Development of Electroplating' [Online]. Available: www.thomasnet.com/articles/custom-manufacturing-fabricating/electroplating-development. [Accessed: 5 June 2017].

[7] R. Clarke, 'DALIC selective brush plating and anodising,' *Int. J. Adhes. Adhes.*, no. 19, pp. 205–207, 1999.

[8] T. Vagramyan, J.S.L. Leach and J.R. Moon, 'On the problems of electrodepositing brass from non-cyanide electrolytes,' *Electrochimica Acta*, vol. 24, no. 2, pp. 231–236, February 1979.

[9] M.R.H. de Almeida, E.P. Barbano, M.F. de Carvalho, I.A. Carlos, J.L.P. Siqueira and L.L. Barbosa, 'Electrodeposition of copper–zinc from an alkaline bath based on EDTA,' *Surf. Coat. Technol.*, vol. 206, no. 1, pp. 95–102, October 2011.

[10] A. He, Q. Liu and D.G. Ivey, 'Electrodeposition of tin: a simple approach,' *J. Mater. Sci. Mater. Electron.*, vol. 19, no. 6, pp. 553–562, June 2008.

[11] A. Collazo, R. Figueroa, X.R. Nóvoa and C. Pérez, 'Electrodeposition of tin from a sulphate bath. An EQCM study,' *Surf. Coat. Technol.*, vol. 280, no. Supplement C, pp. 8–15, October 2015.

[12] D. Fister, 'Reducing operational costs, environmental impact via rigorous plating/finishing analysis,' *Met. Finish.*, vol. 108, no. 6, pp. 39–46, June 2010.

[13] 'IPC-TM-650 Test Methods Manual,' The Institute for Interconnecting and Packaging Electronic Circuits, August 1997.

[14] HSE, 'European Standards and Markings For Respiratory Protection.' January 2013.

[15] 'Waste Classification: Guidance on the classification and assessment of waste – Technical Guidance WM3'. SEPA, NIEA, Natural Resources Wales, Environment Agency, May 2015.

[16] 'How to carry out a COSHH risk assessment' [online]. Available: www.hse.gov.uk/toolbox/harmful/coshh.htm. [Accessed: 9 February 2018].

[17] 'How to comply with your environmental permit Additional guidance for: The Surface Treatment of Metals and Plastics by Electrolytic and Chemical Processes (EPR 2.07)'. Environment Agency, March 2009.

[18] 'Waste legislation and regulations – GOV.UK' [online]. Available: https://www.gov.uk/guidance/waste-legislation-and-regulations. [Accessed: 9 February 2018].

[19] 'Find a local hazardous waste disposal service – GOV.UK' [online]. Available: https://www.gov.uk/hazardous-waste-disposal. [Accessed: 9 February 2018].

[20] Internet Archive Book Images, Image from p. 104 of '*The modern electroplater; a complete book considering fully the elementary principles of electro deposition of metals, their practical application and industrial use*' (1920).

Index

Other Metalworking Guides from Crowood

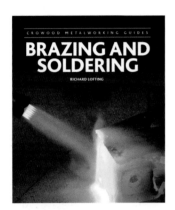

Brazing and Soldering

978 1 84797 836 3

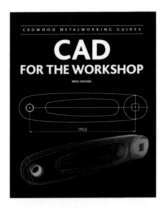

CAD for the Workshop

978 1 84797 566 9

CNC Milling in the Workshop

978 1 84797 512 6

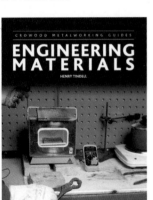

Engineering Materials

978 1 84797 679 6

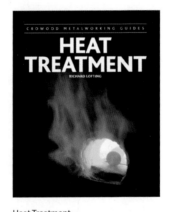

Heat Treatment

978 1 78500 441 4

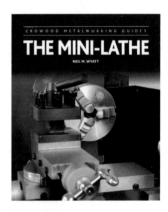

The Mini-Lathe

978 1 78500 128 4

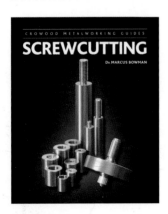

Screwcutting

978 1 84797 999 5

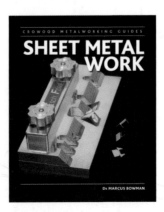

Sheet Metal Work

978 1 84797 778 6

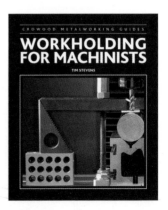

Workholding for Machinists

978 1 78500 238 0